Aging Avionics in Military Aircraft

Committee on Aging Avionics in Military Aircraft

Air Force Science and Technology Board
Division on Engineering and Physical Sciences
National Research Council

NATIONAL ACADEMY PRESS
Washington, D.C.

NATIONAL ACADEMY PRESS **2101 Constitution Avenue, N.W.** **Washington, D.C. 20418**

NOTICE: The project that is the subject of this report was approved by the Governing Board of the National Research Council, whose members are drawn from the councils of the National Academy of Sciences, the National Academy of Engineering, and the Institute of Medicine. The members of the committee responsible for the report were chosen for their special competences and with regard for appropriate balance.

This is a report of work supported by Contract Number F49620-00-1-0174 between the United States Air Force and the National Academy of Sciences. Any opinions, findings, conclusions, or recommendations expressed in this publication are those of the author(s) and do not necessarily reflect the view of the organizations or agencies that provided support for the project.

International Standard Book Number 0-309-07449-5

Limited copies are available from:

Air Force Science and Technology Board
National Research Council
2101 Constitution Avenue, N.W.
Washington, D.C. 20418
(202) 334-3118

Additional copies are available from:

National Academy Press
Box 285
2101 Constitution Ave., N.W.
Washington, D.C. 20055
(800) 624-6242
(202) 334-3313 (in the Washington metropolitan area)
http://www.nas.edu

Copyright 2001 by the National Academy of Sciences. All rights reserved.

Printed in the United States of America

THE NATIONAL ACADEMIES

National Academy of Sciences
National Academy of Engineering
Institute of Medicine
National Research Council

The **National Academy of Sciences** is a private, nonprofit, self-perpetuating society of distinguished scholars engaged in scientific and engineering research, dedicated to the furtherance of science and technology and to their use for the general welfare. Upon the authority of the charter granted to it by the Congress in 1863, the Academy has a mandate that requires it to advise the federal government on scientific and technical matters. Dr. Bruce M. Alberts is president of the National Academy of Sciences.

The **National Academy of Engineering** was established in 1964, under the charter of the National Academy of Sciences, as a parallel organization of outstanding engineers. It is autonomous in its administration and in the selection of its members, sharing with the National Academy of Sciences the responsibility for advising the federal government. The National Academy of Engineering also sponsors engineering programs aimed at meeting national needs, encourages education and research, and recognizes the superior achievements of engineers. Dr. William A. Wulf is president of the National Academy of Engineering.

The **Institute of Medicine** was established in 1970 by the National Academy of Sciences to secure the services of eminent members of appropriate professions in the examination of policy matters pertaining to the health of the public. The Institute acts under the responsibility given to the National Academy of Sciences by its congressional charter to be an adviser to the federal government and, upon its own initiative, to identify issues of medical care, research, and education. Dr. Kenneth I. Shine is president of the Institute of Medicine.

The **National Research Council** was organized by the National Academy of Sciences in 1916 to associate the broad community of science and technology with the Academy's purposes of furthering knowledge and advising the federal government. Functioning in accordance with general policies determined by the Academy, the Council has become the principal operating agency of both the National Academy of Sciences and the National Academy of Engineering in providing services to the government, the public, and the scientific and engineering communities. The Council is administered jointly by both Academies and the Institute of Medicine. Dr. Bruce M. Alberts and Dr. William A. Wulf are chairman and vice chairman, respectively, of the National Research Council.

COMMITTEE ON AGING AVIONICS IN MILITARY AIRCRAFT

ROBERT L. CATTOI, *chair*, consultant, Dallas, Texas
NOEL LONGUEMARE, *vice chair*, Principal Deputy Under Secretary of Defense for Acquisition and Technology (retired), Ellicott City, Maryland
HENRY P. ARNOLD, BF Goodrich Aerospace, Vergennes, Vermont
WILLIAM C. BOWES, Litton Industries, Woodland Hills, California
STEPHEN N. BUSS, Northrop Grumman Corporation, Linthicum, Maryland
JOHN D. COSGROVE, Rockwell Collins (retired), Cedar Rapids, Iowa
FREDERICK H. DILL, IBM T.J. Watson Research Center, Yorktown Heights, New York
LLEWELLYN S. DOUGHERTY, Raytheon Systems Company, El Segundo, California
VALERIE J. GAWRON, Veridian Engineering Flight Research Group, Buffalo, New York
DAVID R. HEEBNER, Heebner Associates, McLean, Virginia
ELLIS F. HITT, Battelle, Columbus, Ohio
ANDREW J. KORNECKI, Embry Riddle Aeronautical University, Daytona Beach, Florida
ROCKY J. PORZIO, Lockheed Martin Federal Systems, Owego, New York
GEORGE W. SUTTON, ANSER, Arlington, Virginia
WILLIAM G.T. TUTTLE, JR., Logistics Management Institute, McLean, Virginia
RAYFORD B. VAUGHN, JR., Mississippi State University, Mississippi
BRIAN T. WRIGHT, Rockwell Collins, Cedar Rapids, Iowa

Liaisons from the Air Force Science and Technology Board

ROBERT A. FUHRMAN, Lockheed Corporation (retired), Pebble Beach, California
ALTON D. ROMIG, Sandia National Laboratories, Albuquerque, New Mexico

Liaisons from the Board on Army Science and Technology

PATRICK F. FLYNN, Cummins Engine Company, Inc. (retired), Columbus, Indiana
GEORGE T. SINGLEY III, Hicks and Associates, Inc., McLean, Virginia

Staff

BRUCE A. BRAUN, Director
JAMES E. KILLIAN, Study Director
GREG EYRING, Consultant
PAMELA A. LEWIS, Senior Project Assistant

AIR FORCE SCIENCE AND TECHNOLOGY BOARD

ROBERT A. FUHRMAN, *chair*, Lockheed Corporation (retired), Pebble Beach, California
ANTHONY J. BURSHNICK, U.S. Air Force (retired), Springfield, Virginia
LYNN CONWAY, University of Michigan, Ann Arbor
WILLIAM H. CRABTREE, consultant, Cincinnati, Ohio
EARL H. DOWELL, Duke University, Durham, North Carolina
ALAN H. EPSTEIN, Massachusetts Institute of Technology, Cambridge
ALFRED B. GSCHWENDTNER, Massachusetts Institute of Technology, Cambridge
ROBERT G. LOEWY, Georgia Institute of Technology School of Aerospace Engineering, Atlanta
JOHN MICHAEL LOH, U.S. Air Force (retired), Williamsburg, Virginia
NOEL LONGUEMARE, Principal Deputy Under Secretary of Defense for Acquisition and Technology (retired), Ellicott City, Maryland
THOMAS S. MOORMAN, JR., U.S. Air Force (retired), McLean, Virginia
BRADFORD W. PARKINSON, Stanford University, Stanford, California
RICHARD R. PAUL, Boeing Company, Seattle, Washington
ELI RESHOTKO, Case Western Reserve University, Cleveland, Ohio
ALTON D. ROMIG, Sandia National Laboratories, Albuquerque, New Mexico

Staff

BRUCE A. BRAUN, Director
MICHAEL A. CLARKE, Associate Director
CHRIS JONES, Financial Associate
WILLIAM E. CAMPBELL, Administrative Coordinator
DEANNA SPARGER, Senior Project Assistant
REBECCA LUCCHESE, Senior Project Assistant

Preface

Questions about the readiness of U.S. military forces to go into battle have received widespread attention in recent years, and the issue was hotly debated in the 2000 U.S. presidential campaign. Partly as a result of a slowdown in purchasing of new materiel in the 1990s, many weapon systems are showing their age, and their lifetimes are being extended beyond their original design lifetimes. This has led to increased maintenance costs and reduced mission-capable rates. The problem is especially severe with electronics systems, such as aircraft avionics, which increasingly depend on commercially available components that have a technology-refresh cycle as short as 18 months and an availability cycle of less than a decade.

This study was requested by the Assistant Secretary of the Air Force for Acquisition to address three areas:

- Provide a "heading check" on what the Air Force is doing to address the aging avionics problem.
- Provide suggestions for additional actions.
- Comment on the division of responsibility between government and industry for maintaining avionics systems.

In the course of its deliberations, the Committee on Aging Avionics in Military Aircraft received many briefings on the aging avionics issue from the perspective of the Air Force, Navy, Army, Office of the Secretary of Defense (OSD), and contractors. The committee also sought out the views of commanders of three Air Force Major Commands, as well as senior executives in the Air Force Secretariat. The committee found that more than 25 different organizations in the military services, OSD, and industry are already working on various aspects of the aging avionics problem. One of the contributions of this study will be to raise the awareness of concerned decision makers of related work being done in various other organizations and to facilitate meaningful coordination among them.

The committee greatly appreciates the support and assistance of National Research Council staff members James Killian, Pamela Lewis, and Carol Arenberg and consultant Greg Eyring in the production of this report.

Robert L. Cattoi, *chair*
Noel Longuemare, *vice chair*
Committee on Aging Avionics in Military Aircraft

Acknowledgments

This report has been reviewed in draft form by individuals chosen for their diverse perspectives and technical expertise, in accordance with procedures approved by the NRC's Report Review Committee. The purpose of this independent review is to provide candid and critical comments that will assist the institution in making its published report as sound as possible and to ensure that the report meets institutional standards for objectivity, evidence, and responsiveness to the study charge. The review comments and draft manuscript remain confidential to protect the integrity of the deliberative process. We wish to thank the following individuals for their review of this report:

Gen. George Babbitt, U.S. Air Force (retired)
Dr. John M. Borky, Tamarac Technologies, LLC
Lt. Gen. Gordon E. Fornell, U.S. Air Force (retired)
Dr. Nancy G. Leveson (NAE), Massachusetts Institute of Technology
Mr. Robert N. Lutter, The Boeing Company
Mr. David W. Mineck, Kaiser Aerospace and Electronics Corporation
Dr. Eberhardt Rechtin (NAE), University of Southern California (retired)

Although the reviewers listed above have provided many constructive comments and suggestions, they were not asked to endorse the conclusions or recommendations nor did they see the final draft of the report before its release. The review of this report was overseen by Raymond S. Colladay, appointed by the Division on Engineering and Physical Sciences, who was responsible for making certain that an independent examination of this report was carried out in accordance with institutional procedures and that all review comments were carefully considered. Responsibility for the final content of this report rests entirely with the authoring committee and the National Research Council.

Contents

Tables and Figures

TABLES

FIGURES

Acronyms

ACAT	acquisition category
AFMC	Air Force Materiel Command
AFRL	Air Force Research Laboratory
AVCOM	avionics component obsolescence management
COTS	commercial off-the-shelf
DMS	diminishing manufacturing sources
DMS/OP	diminishing manufacturing sources/ out-of-production parts
DoD	U.S. Department of Defense
FAA	Federal Aviation Administration
FY	fiscal year
GATM	global air traffic management
GIDEP	Government Industry Data Exchange Program
GIG	Global Information Grid
IPT	integrated product team
IWSM	integrated weapon-system management
JIAWG	Joint Integrated Avionics Working Group
JSF	Joint Strike Fighter
JTA	Joint Technical Architecture
MDAP	major defense acquisition program
Mil Spec	Military Specification
MOSA	modular open-system approach
OASD (C3I)	Office of the Assistant Secretary of Defense (Command, Control, Communications, and Intelligence)
O&M	operations and maintenance
OSD	Office of the Secretary of Defense
PBR	President's Budget Request
QAPR	Quarterly Acquisition Program Review
R&D	research and development
RDT&E	research, development, test, and evaluation
RVSM	reduced vertical-separation minimum
SAF/AQ	Assistant Secretary of the Air Force for Acquisition
SEI	Software Engineering Institute
SPO	system program office
TACTech	Transition Analysis of Component Technology
TCAS	traffic collision-avoidance system
TOC	total ownership cost
USD	Under Secretary of Defense

Executive Summary

BACKGROUND

Since the end of the Cold War, funding for the acquisition of new U.S. military aircraft has become scarce, and after plunging precipitously in the early 1990s, budgets for the modernization of the existing ("legacy") fleet have remained flat. The operational lifetimes of legacy aircraft are being extended well beyond their original design lifetimes, and the average age of U.S. military aircraft is 20 years and increasing. In the 1990s, the U.S. Air Force reported that the mission capability of its aircraft declined by 10 percentage points—from 83 percent to 73 percent. This decline in readiness was due largely to the increasing age of the aircraft fleet, particularly the aging avionics systems upon which the aircraft depend. This trend applies to most military equipment, not just aircraft. As defined in this report, the term "avionics" includes internal electronic hardware, as well as external pods for systems, such as electronic countermeasures. The term "aging" refers to technical obsolescence, as well as physical degeneration over time.

There is widespread agreement that U.S. military forces must be modernized to meet the challenges of the twenty-first century. The critical need to upgrade avionics systems to meet evolving changes in threats, missions, and peacetime air traffic control requirements, especially at a time when very few new programs are being started, should have a high priority. However, the U.S. Department of Defense (DoD) has been caught in a vicious cost spiral that links the costs of modernization with constantly increasing support costs. Because of a relatively flat total budget, the funds needed for modernization are being siphoned off for increasing support costs, which will continue to increase as the equipment ages. This trend must be reversed.

Extending the life of an airframe has proven challenging and costly. Extending the life of an avionics system, however, is one of the most critical and difficult aspects of extending total aircraft system lifetimes. Critical components go out of production or become obsolete, and many former suppliers of military-grade components have gone out of business. From 1986 to 1996, for example, the percentage of discontinued military/aerospace electronic devices nearly doubled—from 7.5 percent to 13.5 percent. In addition, legacy avionics systems, which were designed to meet requirements of the past, generally lack the full capability to perform new missions, meet new threats, or perform well in the new information-intensive battlefield environments.

As the legacy aircraft fleet ages, avionics systems will become more and more difficult to support and maintain. Whereas the military once provided a large and profitable market for the electronics industry, the military electronics market today constitutes less than 1 percent of the commercial market. As a result, the military must increasingly rely on commercial off-the-shelf (COTS) technologies for its avionics hardware and software. Although COTS items are generally less expensive than comparable items designed especially to meet military specifications, the technology-refresh cycle for COTS is typically 18 months or less, which exacerbates the obsolescence problem for aircraft whose lifetimes are measured in decades. The short refresh cycle is driven mostly by the tremendous advances in computer systems, which comprise an increasing percentage of avionics content.

When a new aircraft is designed, the latest advances in avionics technology can be used, and strategies for managing obsolescence can and should be built in from the beginning. However, long weapon-system development and procurement cycles virtually guarantee that some avionics systems will be obsolete by the time they are fielded. The F-22 Raptor program, for example, which began nearly 20 years ago, is still at least five years away from fielding aircraft in squadron strength. The program now budgets $50 million a year to replace "old" avionics with new hardware and software. By the time the first production F-22 rolls off the line, its avionics systems will have undergone four technology-refresh cycles.

According to Lt. General Robert Raggio, Commander of the Aeronautical Systems Center, the Air Force needs an additional $250 million to $275 million per year to address the aging avionics problem in both legacy and new aircraft, not including the cost of training maintenance personnel, suppliers, and operators. Each technology-refresh cycle requires regression testing and flight testing, training for pilots and support personnel, and configuration and spares management, which all add to the implementation cost. Cumulative costs for diminishing manufacturing sources/out-of-production parts (DMS/OP) are projected to reach close to $1 billion each for the F-15, F-22, and U-2.

Without a coherent strategy for managing and containing the total ownership cost (TOC) of avionics systems, both for legacy and new aircraft, the maintenance of these systems will demand an ever-increasing share of the Air Force budget. Managing the DMS/OP problem alone is consuming a larger and larger portion of aircraft program office budgets. If overall DoD budgets remain flat, expenditures on DMS/OP threaten to consume funds that would otherwise be spent on modernizing the aircraft fleet and bringing operational capabilities up to the levels required to counter evolving threats.

STATEMENT OF TASK

In response to a request by the Assistant Secretary of the Air Force for Acquisition, the National Research Council convened the Committee on Aging Avionics in Military Aircraft, under the auspices of the Air Force Science and Technology Board, to conduct this study. The study committee was given the following tasks:

- Gather information from DoD, other government agencies, and industrial sources on the status of, and issues surrounding, the aging avionics problem. This should include briefings from and discussions with senior industry executives and military acquisition and support personnel. A part of this activity should include a review of Air Force Materiel Command's study on diminishing manufacturing sources to recommend ways to mitigate avionics obsolescence.
- Provide recommendations for new approaches and innovative techniques to improve management of aging avionics, with the goal of helping the Air Force to enhance supportability and replacement of aging and obsolescing avionics and minimize associated life cycle costs. Comment on the division of technology responsibility between DoD and industry.

MANAGEMENT OF AVIONICS SYSTEMS

The committee recognizes that there are many dimensions and/or objectives in any strategy for managing the total DMS/OP problem and that individual corrective actions for a particular aircraft platform will depend on the specific characteristics of its installed avionics systems. More than 25 organizations, both inside and outside the Air Force, are working on various aspects of the DMS/OP problem. Although each organization is effective in its limited chartered activities, there is very little coordination among them, and the results of each project are not broadly distributed to the DoD or to the Air Force Enterprise. With a coherent DoD/Air Force strategy for dealing with the DMS/OP problem, collective/coordinated management of these diverse activities could be established, which could result in more productive use of results and minimal redundant expenditures of scarce resources.

A modular open-system approach (MOSA) has been endorsed by the Air Force as a way of developing scalable, more easily upgradable avionics systems and reducing TOCs in both legacy and new aircraft. The committee generally agrees that, if MOSA principles were judiciously applied to new systems and to the updates of legacy platforms, the TOC of avionics systems could be significantly reduced. However, mitigating the aging avionics problem will require more than new technical approaches. Broader issues having to do with DoD management, congressional budgets, and DoD's relationships with its contractors must also be addressed. These issues, as well as the narrower technical issues, are addressed in the following findings and recommendations.

FINDINGS AND RECOMMENDATIONS

Based on testimony from a broad cross section of government managers who are attempting to address the DMS/OP problem for legacy aircraft and new aircraft/avionics subsystems and a broad cross section of prime contractors and subsystem suppliers, who discussed the problem from a private sector viewpoint, the committee arrived at the following findings and recommendations. Note that the committee members were convinced that mitigating the aging avionics problem will require more than new technical approaches and that dealing effectively with the aging avionics problem will not require a technology breakthrough.

General findings are presented first, followed by specific findings in four categories: management issues, budgetary issues, technical issues, and business issues. Because the Air Force is the sponsor of this study, the focus is on actions that should be taken by the Air Force. However, early on during the data-gathering phase of this study, it became readily apparent that the problem is not just internal to the Air Force. All of the services would benefit from a DoD enterprise strategy for dealing with aging avionics. Thus, recommendations are categorized as internal to the Air Force and external to the Air Force. External recommendations should be addressed in a multiservice context at the level of the Office of the Secretary of Defense.

GENERAL FINDINGS

Finding 1. The problem of aging avionics in military aircraft is large and growing. Unless it is addressed proactively and comprehensively, it will have a significant negative impact on the military readiness of U.S. forces.

Finding 2. The amount budgeted for the modernization of Air Force avionics systems is far short of the amount needed to pay for upgrades already approved in critical areas: performance and safety-mandated upgrades; avionics upgrades required for the global air traffic management (GATM) system; and replacements for aging avionics subsystems with the lowest reliability and/or highest repair costs.

Finding 3. A large number of organizations within DoD, the military services, and industry are attempting to address various aspects of the aging avionics problem. However, these efforts are poorly coordinated and often duplicative.

Finding 4. Widespread application of a MOSA to avionics architectures would enable DoD to manage the aging avionics problem more affordably, for both new aircraft and many legacy systems.

Finding 5. Most of the benefits of MOSA can be realized through a "modular" approach. Although a fully "open" system would have some additional advantages to the government in a few situations (as they do in certain commercial sectors where quantities and related factors can support a viable business case for this approach), most DoD acquisitions cannot justify a totally open approach. The "modular" aspect of MOSA, however, could be applied to virtually all DoD products.

SPECIFIC FINDINGS IN KEY AREAS

Government Management

Finding 6. There is no DoD-wide enterprise strategy, and only an embryonic Air Force-wide strategy, for dealing with the aging/obsolescent avionics problem. As a result, no enterprise management or leadership is addressing the problem on a full-time basis.

Finding 7. The Joint Technical Architecture (JTA) for defining weapon system architectures and standards extends beyond those needed for *inter*platform interoperability. The extension into *intra*platform standards is neither consistent nor integrated with MOSA approaches for addressing aging avionics. In fact, the JTA has shown an alarming reversion to the Military Specification (Mil Spec) era by requiring an onerous number of standards and specifications for *intra*-platform avionics systems.

Finding 8. The technical expertise of DoD's depot support maintenance personnel in state-of-the-art avionics systems appears to be eroding as the workforce ages and retires.

Finding 9. As modifications and upgrades of aging avionics systems continue, aircraft, even of the same type, are being equipped with avionics systems with different compositions, capabilities and compatibilities, thus exacerbating the configuration-management problem.

Budgetary Issues

Finding 10. Long acquisition and upgrade cycles virtually require that avionics technology-refresh cycles be built into program plans during the engineering and manufacturing development phase prior to initial fielding.

Finding 11. Because of legal restrictions on the use of appropriated funds in various segregated accounts ("colors of money"), program managers are unable to address aging avionics problems in the most efficient way.

Finding 12. A comprehensive MOSA solution to the aging avionics problem could save money in the long run but would generally cost more than customized point solutions in the short run. This is particularly true for avionics upgrades in the legacy fleet.

Technical Issues

Finding 13. Implementation of MOSA would be facilitated by addressing the following needs:

- development of a common understanding of MOSA
- support for development of MOSA building codes, and disciplined design processes and related design tools required for MOSA implementation
- development of a test/requalification strategy coupled with the proper modeling and simulation tools to implement the MOSA strategy economically
- development of design-reuse databases and high-fidelity avionics models by original equipment manufacturers and suppliers

Business Issues

Finding 14. MOSA challenges the traditional military procurement model in several ways:

- With a modular, open-structured avionics system, DoD would, in theory, be able to solicit supplier competition at a variety of systems architecture levels: at the component level, the circuit-board level, the module level, or the subsystem level. The level must be high enough to provide incentives for qualified suppliers to participate, take advantage of local openness, and encourage suppliers to invest in research to improve avionics systems and stimulate innovation.
- The traditional mind-set of acquiring hardware and software will have to be changed to one of acquiring functionality (an approach in keeping with acquisition-reform precepts).
- The protection and value pricing of a supplier's intellectual property will be a key to success and will therefore require workable business models.
- Business incentives must be defined and provided to suppliers that will motivate a MOSA to avionics system design.

Finding 15. As DoD relies more heavily on commercial off-the-shelf hardware and software in avionics systems—and less on Mil Spec components and DoD-unique software languages—the expertise and intellectual property necessary to develop and maintain these systems will increasingly reside in the private sector.

RECOMMENDATIONS

Recommendations Specific to the Air Force

Recommendation 1. The Air Force, in coordination with the Office of the Secretary of Defense, should develop an "enterprise strategy" for dealing with the aging avionics problem. As a central feature of this strategy, the Air Force should mandate the creation of platform management/upgrade road maps with defined funding requirements for each weapon-system program.

Recommendation 2. The Air Force should raise the awareness in Congress about the shortfall in funding for avionics modernization by increasing its congressional budget request to a level consistent with the modernization plans in system road maps.

Recommendation 3. The Air Force should require a modular, open-system design strategy for all new programs and upgrades, unless specifically waived. Emphasis should be on achieving the benefits of modularity rather than on complete openness, which often creates business or technical problems. A training program in MOSA concepts should be included for program managers, acquisition personnel, and support personnel. Contractors should be encouraged to use executable specifications as the primary archival documentation of the system; these specifications should be integrated into the avionics design environment.

Recommendation 4. The Air Force should continue to use the Quarterly Acquisition Program Reviews (QAPRs) as a forum for top-level oversight and, most important, for setting priorities to address the aging avionics problem.

Recommendation 5. The Air Force software and hardware testing community should develop a testing/requalification strategy tailored to modular avionics systems and should explore methods, including the use of high-fidelity simulation/emulation models and test beds, to minimize the impact on cost and schedule of requalifying avionics components and systems. The Air Force should build on the test strategy and simulation/emulation/diagnostic software model used by the Federal Aviation Administration in the commercial sector, which recognizes the value of reusing hardware/software and provides certification-test credit for reusable modules.

Recommendation 6. The Air Force should examine the feasibility of requiring, as a normal contractual deliverable, contractor-retained high-fidelity avionics simulation models as a means of minimizing validation/certification testing.

Recommendation 7. The Air Force should increase its support for the new Aging Aircraft System Program Office (SPO), in the Aeronautical Systems Center (ASC), by reinforcing its leadership and management responsibility for reducing the total ownership costs of new and legacy avionics systems.

Recommendation 8. The Air Force should develop and apply innovative contracting approaches that provide incentives for both government and contractors to reduce total ownership costs of avionics systems.

Recommendations That Apply to All of the Services

Recommendation 9. The Air Force should recommend that the Office of the Secretary of Defense develop an overall "enterprise strategy" for dealing with the aging avionics problem and issue a specific policy directive covering the following four points:

- A modular, open-system design strategy should be required for all new programs and upgrades, unless specifically waived.
- Development and use of program road maps should be mandatory for all Acquisition Category I (ACAT-I) programs (and their use encouraged for lesser programs); road maps should include funding plans and anticipated reductions in total ownership costs.
- Reviews by the Defense Acquisition Board (DAB) of these items should be a required acquisition milestone exit criteria.
- A revolving fund should be established (possibly the Working Capital Fund) to further front-end design/qualification of MOSA-compatible solutions to the problem of diminishing manufacturing sources.

Recommendation 10. The Air Force should recommend that OSD form joint working groups with industry to address policy and business concerns involved in the resolution of aging avionics problems:

- An industry/government steering group should be formed as a focal point for addressing the issues raised by MOSA procurement models and related modifications to the acquisition process, business/competitive models, intellectual property rights, management/pricing, the 50/50 rule, and related issues.
- The role of the Software Engineering Institute (SEI) could be expanded to include the development of MOSA building codes and design tools and processes; SEI could also recommend the process for defining and implementing interface standards at the proper point in the design cycle. The committee believes these changes would be consistent with current plans to reorganize SEI to consolidate software development, system

development, and integrated product team (IPT) activities.

- Congress should be encouraged to give DoD managers greater flexibility to shift funds among budget categories to take advantage of opportunities to reduce total ownership costs (TOCs).
- DoD should consider avenues to encourage young people to seek engineering educations focused on embedded software intensive systems and the maintenance of legacy systems.

Recommendation 11. The Air Force should recommend that OSD form a joint cross-platform working group (JCWG) at the flag-officer level to focus on reducing total ownership cost through the joint development of modular, scalable systems and the use of common solutions across weapon system platforms.

Recommendation 12. The Air Force should recommend that OSD examine and modify traditional defense procurement practices to minimize problems for avionics suppliers.

Recommendation 13. The Air Force should recommend that the current Open Systems Joint Task Force become the center of expertise and the focal point for addressing issues associated with the application of MOSA. Modularity, rather than total openness, should be emphasized to accommodate current business and technical issues.

Recommendation 14. The Air Force should recommend that the Office of the Under Secretary of Defense for Acquisition, Technology and Logistics restrict applicability of the Joint Technical Architecture (JTA) and mandated standards to *inter*platform interoperability and allow the *intra*platform standards to be defined by a MOSA approach, along with a greatly reduced number of mandated standards.

1

Introduction

Since the end of the Cold War, funding for the acquisition of new military aircraft has become scarce, and budgets for modernizing the existing, so-called "legacy" fleet have remained flat. As a result, the operational lifetimes of legacy aircraft are being extended well beyond their original design lifetimes. The average age of U.S. military aircraft is 20 years and increasing as a result of the low replacement rate. Figure 1-1 shows the almost year-by-year increase in the age of aircraft since 1997. Although extending the lifetime of the airframe is relatively straightforward, avionics systems, which are often based on technology from the 1970s and 1980s, are rapidly becoming obsolete. Even if these systems could be adequately maintained, they are generally not adequate for dealing with current and evolving missions, threats, and information-intensive battlefield environments.

As legacy aircraft age, the avionics systems are becoming more and more difficult to support and maintain. Many critical components are no longer in production or have become obsolete, and many former suppliers of military-grade components have either gone out of business or have stopped producing for the military market. Thus, more and more aircraft are being grounded while maintenance and support solutions are pursued. The Air Force reports that the mission-capable rate (i.e., the percentage of aircraft able to perform their primary missions, at any given time) of its aircraft declined from 83 percent to 73 percent during the 1990s, and indications are that this trend will continue in the near future (CBO, 2000). The Air Force attributes this decline in readiness largely to the aging of the aircraft fleet, particularly the aging of avionics[1] systems upon which the aircraft depend (personal communication with General John Jumper, Commander, Air Combat Command, August 4, 2000). The term "aging," usually used to refer to the degeneration and failure of components over time, is used in this report to refer to technical obsolescence in addition to physical degeneration.

Not long ago, the military provided a large and profitable market for the electronics industry. Since 1995 the military market has constituted less than 1 percent of the commercial integrated circuit market (Figure 1-2). As a result, the military must rely increasingly on commercial off-the-shelf (COTS) technologies[2] for both avionics system upgrades and new designs. Although COTS items are generally less expensive than comparable items designed to military specifications, the

[1] As defined in this report, the term "avionics" includes: internal electronic hardware, as well as external pods, such as electronic countermeasures; software required for navigation, communication, and other functions; external automatic test system hardware and software; ground electronics, communications, and air traffic control hardware.

[2] The term COTS is used here to mean any developed commercial technology available for sale; it need not be mass produced.

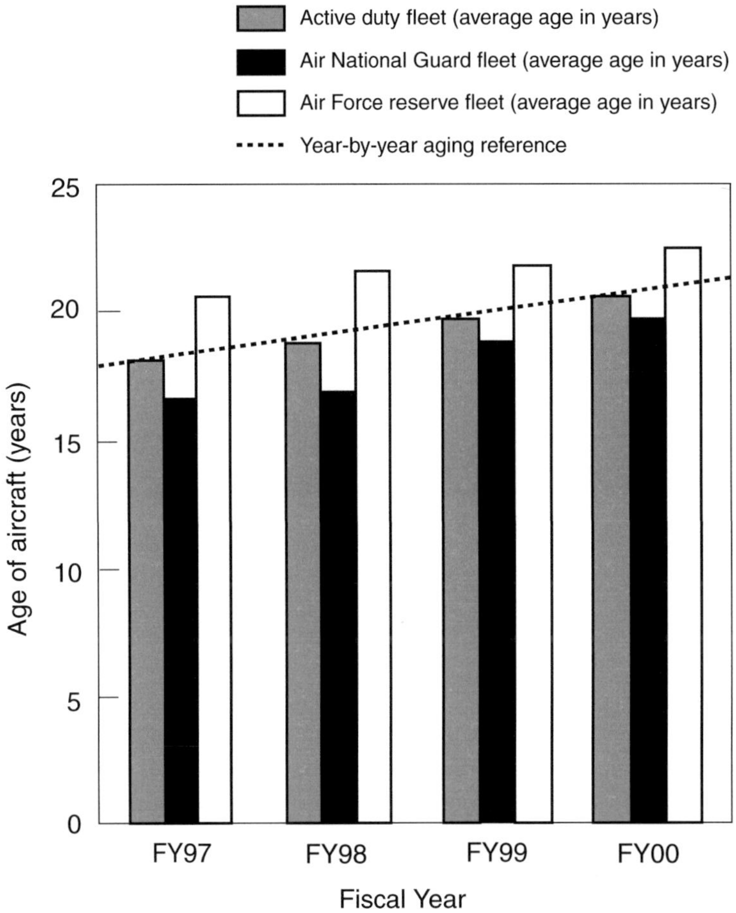

FIGURE 1-1 Average age of U.S. Air Force aircraft. Source: U.S. Air Force, 2000a.

technology-refresh cycle for COTS is typically 18 months or less, which exacerbates the ongoing problem of obsolescence for aircraft with lifetimes measured in decades.

Long weapon-system development and procurement cycles are also part of the problem. The F-22 Raptor program, for example, was begun nearly 20 years ago and is still at least five years away from fielding aircraft in squadron strength. Currently, $50 million a year is being budgeted to replace the "old" F-22 avionics with new hardware and software (Raggio, 2000). By the time the first production F-22 rolls off the line, its avionics systems will have undergone four refresh cycles.

According to Lt. General Robert Raggio, Commander of the Aeronautical Systems Center, the Air Force needs an additional $250 million to $275 million per year to address the problem of aging avionics in both legacy and new aircraft, not including the costs of training maintenance workers, suppliers, and operators (personal communication with Lt. Gen. Robert Raggio, Commander, Aeronautical Systems Center, October 6, 2000).[3] Each technology-refresh cycle entails added costs for regression testing, flight testing, training for pilots and support personnel, and configuration and spares management.[4]

In the 1980s, the Joint Integrated Avionics Working Group (JIAWG) was formed to establish a set of avionics characteristics for all of the services and for multiple platforms. Three aircraft were selected for initial application of the JIAWG principle: the Air Force advanced technology fighter (now the F-22); the Navy A-12 fighter; and the Army Comanche helicopter. The JIAWG also developed hardware standards, including

[3] Training costs for design and test engineers, logisticians, maintenance personnel, and aircrews, etc., are not currently included in cost models for aging avionics.

[4] No institutionalized processes, tools, or requirements have been developed for configuration management.

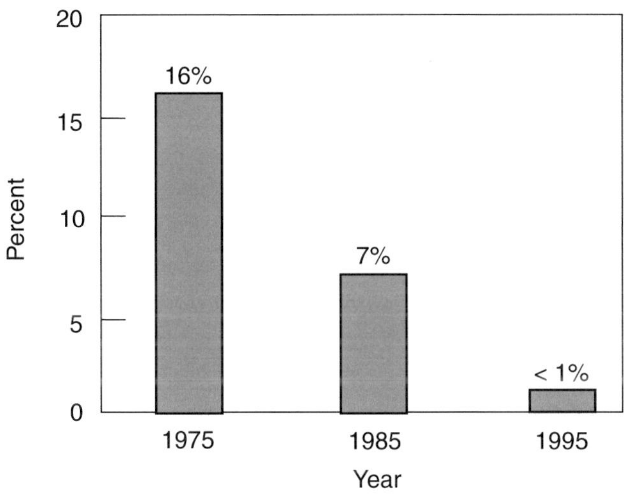

FIGURE 1-2 Decline in the military market share for integrated circuits. Source: Courtesy of Aerospace Corporation.

contractor-unique interfaces and bus structures, and the U.S. Department of Defense (DoD) mandated that Ada (a computer language) be used for software. In the committee's opinion, JIAWG's goals (i.e., reducing development, production, and support costs through the use of common items for all services) were laudable. In retrospect, however, the effort was fundamentally flawed because JIAWG's basic strategy was to define common modules, on the assumption that architecture would naturally flow from the module catalog. However, this approach is contrary to good system engineering (personal communication with Dr. J.M. Borky, Chief Scientist, Tamarac Technologies, February 16, 2001). In addition, the JIAWG did not anticipate the explosion in commercial electronics or the enormous market driving it. As a consequence, the system recommended by JIAWG proved to be unaffordable because it was predicated on government-supported research and development (R&D) and government markets to sustain the product lines.

Since the 1980s, the Air Force and the other military services have commissioned numerous task forces and committees and have funded programs to address various aspects of the aging/obsolescent avionics problem. Many of these have been successful in a relatively narrow or channeled area. However, these endeavors were not coordinated, were sometimes redundant, and, in general, did not use resources (human and monetary) synergistically, or develop and apply best practices.

In addition, the Office of the Secretary of Defense (OSD) and the Air Force have endorsed a modular, open-system approach (MOSA)[5] to ensure that new avionics systems will be more extendable and easily upgradable, as well as to reduce total ownership costs (TOC)[6] and improve readiness. All of the military services are beginning to recognize that MOSA could also result in significant benefits for upgrading and retrofitting to other types of systems. A major purpose of this study is to evaluate this approach.

STATEMENT OF TASK

This study was requested by the Assistant Secretary of the Air Force for Acquisition (SAF/AQ).[7] The study Committee on Aging Avionics in Military Aircraft, established by the National Research Council, was asked to perform the following tasks:

- Gather information from DoD, other government agencies, and industrial sources on the status of, and issues surrounding, the aging avionics problem. This should include briefings from and discussions with senior industry executives and military acquisition and support personnel. A part of this activity should include a review of Air Force Materiel Command's study on diminishing manufacturing sources to recommend ways to mitigate avionics obsolescence.
- Provide recommendations for new approaches and innovative techniques to improve management of aging avionics, with the goal of helping

[5] "Modular" systems involve the isolation of functional performance from the specific characteristics of the hardware and software used to implement the system. Ideally, an obsolete part could be removed and substituted with an upgrade without affecting the characteristics of the rest of the system. "Open" systems are generally modular but make use of nonproprietary interface definitions and standards available to multiple competitors. In theory, several prospective suppliers of an avionics module could compete for production and maintenance contracts, thus lowering the acquisition costs to the government.

[6] Total ownership costs are costs incurred over the entire life cycle of an avionics system, including research and development, manufacturing, and maintenance of the system throughout its service life.

[7] Because the Air Force requested the study, the majority of presentations made to the committee and much of the data gathered relate specifically to the Air Force. However, the committee notes that the problem of aging avionics is common to all of the services and that many of the solutions discussed will require a departmentwide approach to be effective.

the Air Force to enhance supportability and replacement of aging and obsolescing avionics and minimize associated life cycle costs. Comment on the division of technology responsibility between DoD and industry.

FORMS OF OBSOLESCENCE

Military equipment "ages" in two basic ways: obsolescence in hardware or software that renders the equipment insupportable; and inadequate performance that renders the equipment unable to fulfill its mission. These problems are most severe in legacy aircraft but are also encountered in new systems, unless steps are taken to preclude or mitigate the problem.

Obsolescence of Hardware and Software

Aging Hardware

Legacy aircraft and electronics in general are both less reliable and more expensive to maintain than for new aircraft. The decrease in reliability is often attributable to the use of discrete, analog parts rather than integrated, digital components. In addition to the inherent differences in the reliability of parts, systems are also less reliable because of the lack of robustness in functional designs associated with cumbersome analog design processes. When older parts go out of production, (i.e., when the manufacturer no longer produces the units on a regular basis—or at all) numerous complications arise: an inventory of spare parts may not exist; the supplier is faced with the high cost of restarting the production line; and subtier manufacturers may have disappeared. In the worst cases, when no suitable components can be found, a redesign becomes necessary, which raises new problems: longer time to fill an order and the commitment of valuable engineering resources for a low-volume redesign (with attendant low profit margins). Thus, both the government and the manufacturer are in a losing situation (Hitt, 2000).

There are several ways of coping with diminishing manufacturing sources/out-of-production parts (DMS/OP). The three main ways are:

- purchase a lifetime supply, with attendant inventory costs
- redesign circuits to accept different, available parts or emulate the functionality provided by an obsolete part using newer technology
- replace entire modules or subsystems with new technology

Each of these strategies may be appropriate depending on many factors, such as the remaining service life of the particular platform or system, cost trade-offs, available budgets, and so on. However, coping with DMS/OP is more difficult than it first appears. Legacy avionics systems were not designed for ease of change or ease of testing. Therefore, the costs and complexity of inserting even new, available components can be high.

The often lengthy regression and flight testing required to validate that changes have not adversely affected safety or overall system performance have an even greater impact. Because of the structure of legacy avionics architectures, which have historically involved numerous, often subtle interactions between disparate components of a system, the causes and effects of changes are difficult to understand and even more difficult to predict. Therefore, extensive testing of the resulting system must be done to verify that no unforeseen consequences are lurking in the background. This process can be very time consuming and expensive.

Commercial Off-the-Shelf Technology

The defense industry is increasingly using COTS hardware and software, which raises another set of problems. In general, the half-lives of commercial electronic products are much shorter than those of military platforms. Although backward compatibility is sometimes possible, commercial business strategies are often based on planned obsolescence, which virtually guarantees that a COTS product will undergo several changes during the lifetime of any piece of military equipment. Therefore, a strategy for dealing with the obsolescence of COTS products must be developed.

Aging Software

Software obsolescence is a growing problem, especially for legacy equipment. Although software itself does not wear out, it must often be modified to accommodate incremental improvements and changes required to implement periodic block upgrades. Avionics software is developed in a host-target paradigm. The software is written, translated, simulated, and verified on a general purpose workstation with a rich development environment. The resulting object code is then downloaded for further testing and validation to the embedded target computer. The rapid progress of computing technology has rendered many host workstations obsolete and forced industry to replace them with new machines (not many PCs nowadays run on 16 MHz processors or boot to DOS). While the hardware is being upgraded, tools developers update their products (by discarding them and not supporting earlier versions). The updated versions of development tools (e.g., compilers, debuggers, linkers, and simulators) are often not compatible with the existing target software, requiring significant development efforts to produce a working product even when the original source code is available. Such scenarios have forced the industry to introduce frequent target software changes, as well as system redevelopment.

Many legacy military platforms contain software written in a variety of obsolete or obsolescing languages, such as machine-assembly languages, JOVIAL, and, to a lesser extent, Ada.

The commercial market, for which the vast bulk of software is written, has evolved its own languages (e.g., C, C++, JAVA, etc.), and funding for most R&D on software and related support tools is now directed toward supporting these languages. Because most software courses taught by U.S. educational institutions are focused on the needs of the commercial world, the number of software engineers skilled in legacy, military-unique languages is shrinking. Thus, the obsolescence of military software is complicated by two problems: (1) the increasing cost of maintaining legacy software maintenance tools; and (2) the decreasing number of technical personnel experienced in legacy software. In addition, much of the documentation for legacy software is inadequate by today's standards and can only be interpreted by specialized personnel.

Inadequate Performance of Hardware/Software Systems

Another type of avionics systems obsolescence is inadequate performance of aging systems in terms of meeting internal or external requirements of the related weapons systems.

Internal Performance Requirements

Internal requirements encompass improvements in safety, reliability, and maintainability, which are usually mutually dependent. Systems upgrades that address all three internal requirements reduce maintenance costs and increase availability and readiness. Orders-of-magnitude improvement in reliability can only be achieved through advanced, solid-state electronics and disciplined design processes.

External Performance Requirements

System upgrades for external requirements include the ability to fulfill new missions, meet new threats, and operate in the evolving global air traffic control system. Sensor systems, such as radar, must cope with increasingly small target cross sections and growing surveillance requirements; avionics must be modified to be compatible with new precision weapon systems; and electronic warfare systems must be continually updated to meet new threats. Because most of these requirements increase demands on legacy computing/processing capabilities, new hardware, as well as significant changes in software, are required.

New warfighting priorities, which revolve around information superiority, also have a large impact on military electronic systems. Battlefield strategies increasingly call for a "system-of-systems" approach, in which assets on land, sea, and air must be interoperable and closely coordinated. This requires capabilities for high-speed data transmission and processing, including the ability to receive and process information from anywhere in the operational network. These capabilities will require significant upgrades to avionics systems on a continual basis.

The avionics content of new airframes, such as the Joint Strike Fighter (JSF), is approaching 40 percent of total system cost because of the greatly expanding modes and features now expected of modern fighter aircraft and the increasing use of multiplatform, off-board information (see Figure 1-3). Even though legacy aircraft cannot possibly match these capabilities, they must be upgraded as much as practical, especially if system interoperability is required.

Required Upgrades in a Free-Flight Environment

Many existing types of aircraft need mandatory avionics upgrades to operate in the air traffic environment. For example, avionics dictated by the Global Air Traffic Management (GATM) Program will have to be installed in most of the Air Mobility Command aircraft, many of which are already undergoing mandated

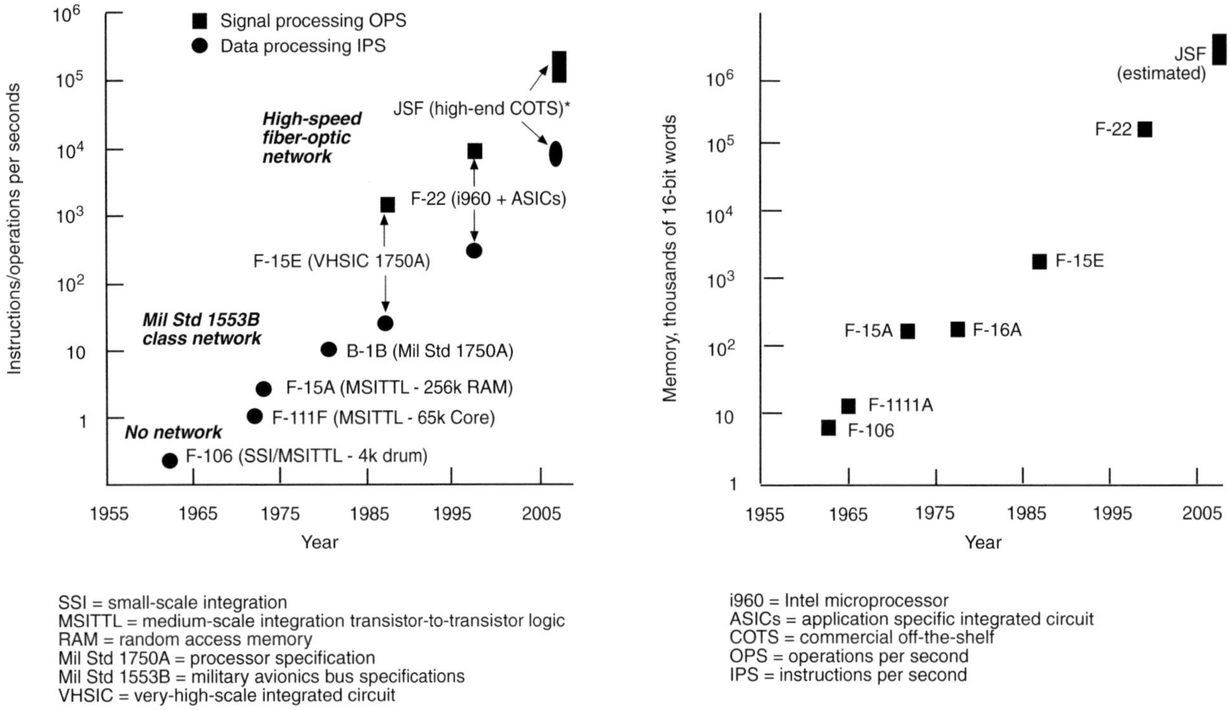

FIGURE 1-3 Historic trends in avionics processing.

modifications, including the installation of the traffic collision avoidance system (TCAS) and ground proximity warning equipment. Combat aircraft from Air Combat Command may be required to make similar modifications in the future. These modifications do not address problems presented by avionics systems/subsystems that drive high maintenance costs and high TOCs.

The modernization/upgrading of avionics systems and the support of older, out-of-production components are related. Indeed, mandatory modernization may provide opportunities to address the DMS/OP problems. The capability/performance upgrades necessary for aircraft platforms to operate for long service lives could be combined with other improvements, such as replacing older avionics components that have particularly high support costs, for relatively little additional cost.

FUTURE MANAGEMENT OF OBSOLESCENCE

In the future, the problem of aging avionics must be addressed in the design of current and new systems. The bottom-line goal of MOSA is to alleviate the DMS/OP maintenance problem and to accommodate modifications and upgrades economically, thus reducing TOCs and improving readiness. Mitigating the aging avionics problem will require more than new technical approaches, but no technology breakthroughs will be necessary. The challenges can be grouped into four categories:

- *Enterprise Management.* DoD and the Air Force are complex organizations with fragmented management responsibilities for weapon system platforms. Implementing common solutions across various vintages of a single platform, different platforms, and across the services is extremely difficult.
- *Budgetary Challenges.* Managers faced with flat or declining discretionary budgets often lack the resources and flexibility to replace avionics components and subsystems that have high operating costs with designs with lower TOCs.
- *Technical Challenges.* The goal of MOSA is to make avionics systems easier to change and

upgrade; however, the concept has not yet been fully defined, and its acceptance as a DoD-wide design and development strategy for avionics will require better supporting tools and retraining of personnel.

- *Business Challenges.* DoD and industry will have to develop new business models that support competition and investment in R&D by suppliers in a MOSA environment. Business incentives must be defined and included in the avionics acquisition process.

REPORT STRUCTURE

Chapter 2 is a broad overview of the magnitude of the problem of aging avionics. Chapter 3 provides an overview of ongoing government and industry initiatives to address the problem of aging avionics. In Chapter 4, the committee presents its observations and assessments based on analyses of presentations, briefings, and data from other sources. Chapter 5 provides a summary of the committee's findings and recommendations.

2

Magnitude of the Problem

When the Cold War ended, it became possible to reduce the size of U.S. forces significantly. Between 1989 and 1999, the number of active duty military personnel, as well as civilian DoD employees, was reduced by 34 percent (CBO, 2000). Because of the downsizing, a surplus of equipment became available from the procurement programs of the 1980s; therefore, there was a corresponding reduction in the purchasing of new weapons.

By the end of the 1990s, the downsizing was essentially complete. However, because of the downturn in procurement, the average age of many kinds of military equipment had increased. This older equipment requires increased maintenance and is vulnerable to a lack of parts, which has led to the cannibalization of one unit to keep another running. The overall result has been lower mission-capable rates and a decrease in readiness.

Although U.S. military forces must be modernized to meet the challenges of the twenty-first century, DoD has been caught in a vicious cost spiral of modernization costs and constantly increasing support costs. Because of a relatively flat total budget, funds needed for modernization are often siphoned off to meet growing support costs, which continue to increase as equipment ages. This trend must be reversed. The problem of maintaining and modernizing aging avionics is acute.

DIMINISHING MANUFACTURING SOURCES/ OUT-OF-PRODUCTION PARTS

As the size of U.S. forces has decreased, there has been a corresponding consolidation of the defense industrial base, including a consolidation of the suppliers of avionics components. The reduction in the number of prime contractors, combined with reduced procurement budgets, has led to a commensurate reduction in market opportunities for lower tier suppliers. This has further exacerbated the DMS problem.

Even companies that have continued to supply DoD have, in many cases, shifted their focus to meeting the requirements of commercial markets, which are characterized by ever shorter product life cycles. As a result of these trends, fewer suppliers of legacy avionics components are available today, and parts that are available are going out of production at an accelerating pace.

Transition Analysis of Component Technology (TACTech), Incorporated, is a company that tracks the availability of electronic components and provides information tools for managing parts obsolescence. Table 2-1 shows the rates at which standardized military/aerospace devices listed in the TACTech database went out of production between 1986 and 1996. During that 10-year period, the percentage of parts that

TABLE 2-1 Accelerating Obsolescence of Military/Aerospace Devices

Year	Number of Parts in TACTech Database	Number of Parts Discontinued	Percentage of Parts Discontinued
1986	22,341	1,675	7.5
1988	30,811	2,975	9.6
1990	55,326	4,371	7.9
1992	72,089	7,593	10.9
1994	58,295	9,659	16.5
1996	45,873	6,210	13.5

Source: TACTech, 1997.

were discontinued almost doubled, from 7.5 percent of the total to 13.5 percent of the total. There is every reason to believe that these percentages will increase in the future. Although the total number of unique part styles is decreasing as levels of circuit integration increase, the percentage of discontinued parts is not expected to decrease at the same rate.

A significant portion of funds allocated to each weapon system is being used to contend with the DMS/OP problem. Estimates of the cumulative amount of money required to address DMS/OP for the F-15, F-22, and U-2 (including development, production, and installation) are close to $1 billion each (U.S. Air Force, 2000a). It is important to stress that these funds are required simply to maintain current functions and do not buy any additional capability.

RISING SUPPORT COSTS

A DoD report, *Product Support for the 21st Century: A Year Later* (September 2000) notes that DoD spends about $62 billion annually to support and maintain its equipment (DUSD[AT&L], 2000). In fiscal year 1999 (FY99), the Air Force spent about $3 billion for depot-level repairs of its aircraft. Approximately one-third of this, or $1 billion, was spent on maintaining and supporting avionics systems (operations and maintenance [O&M] funds), as shown in Figure 2-1. An additional

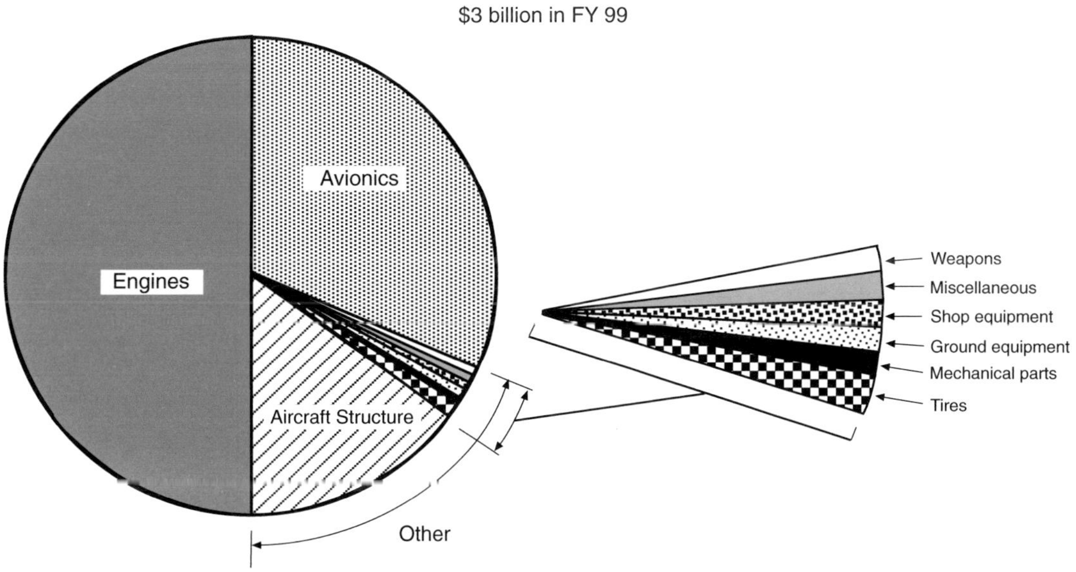

FIGURE 2-1 Cost of avionics in depot-level aircraft maintenance for FY99. Source: U.S. Air Force, 2000a.

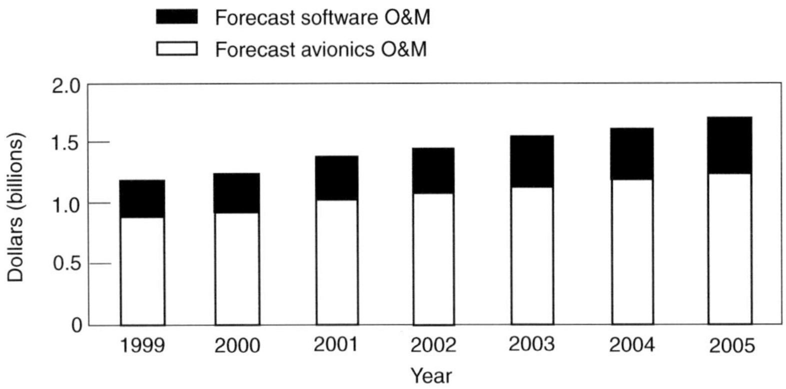

FIGURE 2-2 Projected depot-level avionics operations and maintenance costs. Source: U.S. Air Force, 2000b.

$250 million to $275 million per year is needed to address the aging avionics problem (personal communication from Lt. Gen. R. Raggio, Commander, Aeronautical Systems Center, July 2, 2000). In fact, avionics systems are the second largest component of Air Force O&M costs after engines.

Because of the growing DMS/OP problem, depot-level support costs for avionics are projected to increase by about 50 percent in the next five years (Figure 2-2). Monies spent strictly on DMS/OP for one-for-one replacement are not available for modifications that could improve the reliability or maintainability of avionics components or reduce TOC. Thus, the Air Force is compelled to continue to play catch-up with its O&M funds. For example, O&M budget constraints on the Air Mobility Command have left insufficient funds for the C-5, C-141, and C-17 to fly the required number of flight hours to meet training requirements (U.S. Air Force, 2000a).

BUDGET FOR MODERNIZING AVIONICS

The Air Force must maintain an inventory of approximately 6,000 aircraft to sustain 195 active air wings. For the past five years, the average annual procurement of new aircraft has been only 25 aircraft: 22 in 1996, 22 in 1997, 24 in 1998, 26 in 1999, and 28 in 2000 (Hitt, 2000). If this low rate of procurement continues, the USAF will turn over its aircraft inventory every 240 years. Until something is done to reverse this trend, as the age of aircraft increases, O&M costs will also increase. With an essentially flat DoD budget and strong pressure against increasing aircraft procurements, O&M dollars are being diverted from budgets for modernization, which exacerbates the problem. The limited remaining dollars for modifications are being used to fund modifications to enable airplanes to operate in controlled airspace and to make existing aircraft compatible with the new "smart weapons" that are coming into the inventory. Consequently, very few dollars are left to modernize aging avionics systems or the infrastructure to support these systems.

The Air Force modernization account (modernization includes R&D, testing, evaluation, and procurement), approximately $20 billion per year, has remained at about that level throughout the 1990s and is projected to remain flat through FY07 (Durante, 2000). Figure 2-3 shows funding for avionics modernization from the FY01 President's Budget Request (PBR) (PBR, 2000).

As Figure 2-3 shows, funding would increase through FY01 and decrease thereafter. According to the committee's analysis, however, the avionics investments already approved in the FY01 PBR will cost an additional $5 billion beyond FY05, which is inconsistent with the decrease after FY01. Figure 2-4 shows a breakdown of funds already committed to out-year costs by weapon system, which are dominated by modifications to the C-130.

Some of the upgrades funded in the PBR will be necessary to ensure that U.S. aircraft meet the requirements of the GATM. In addition, most of the transport aircraft from Air Mobility Command will be provided with the TCAS and ground proximity warning equipment. These upgrades account for approximately 20 percent of the modernization budget each year.

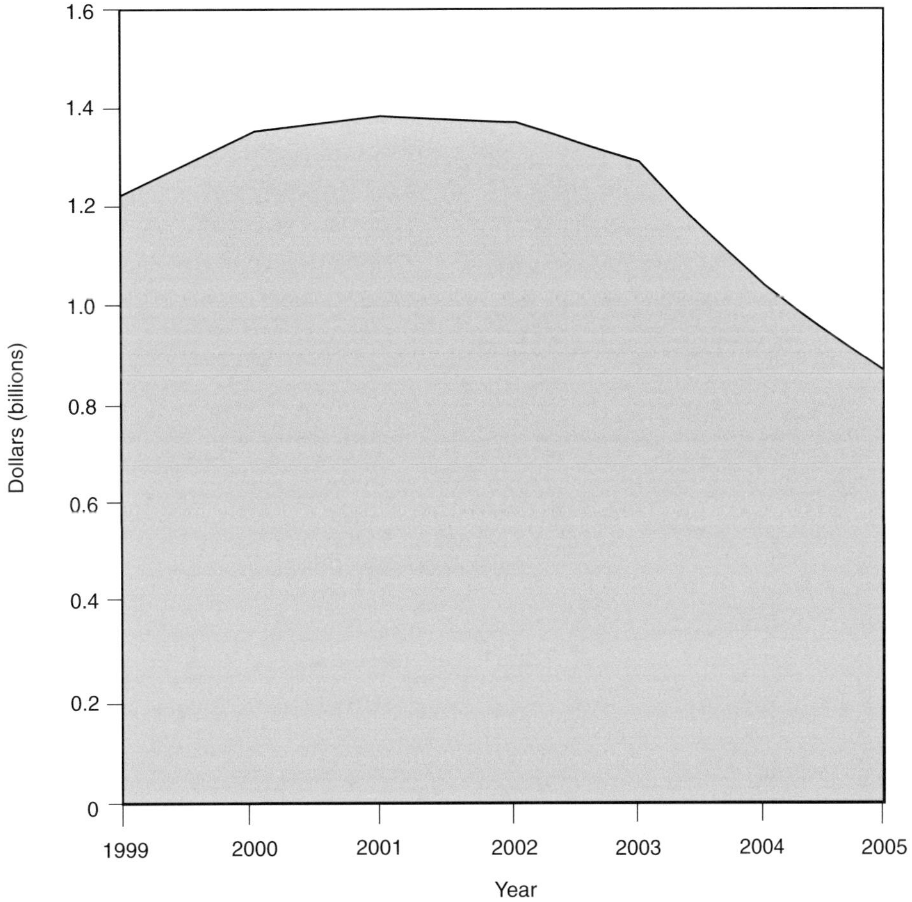

FIGURE 2-3 FY01 President's Budget Request for avionics modernization. Source: PBR, 2000.

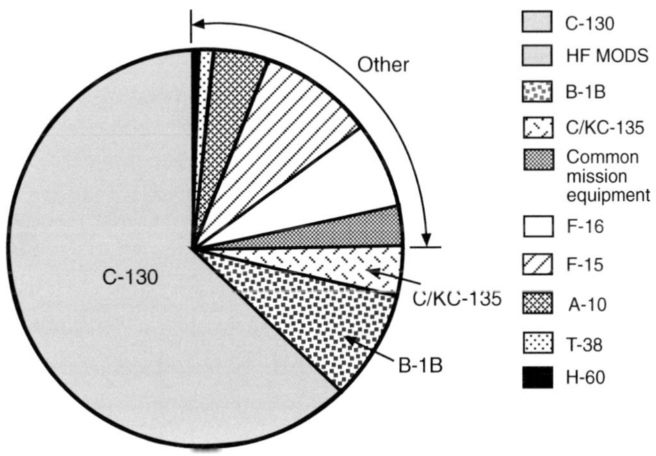

FIGURE 2-4 Out-year costs after FY05 for avionics modernization (approximately $5 billion). Source: U.S. Air Force, 2000a.

Unfortunately, these modifications do not replace high-TOC subsystems or out-of-production avionics components and, therefore, will not substantially improve the DMS/OP picture. For instance, the complex F-16 APG-68 radar has the highest O&M cost of all F-16 avionics, yet none of the planned modifications in the budget involve upgrading or replacing the F-16 radar (PBR, 2000). Because GATM upgrades are considered necessary for aircraft to continue flying, they take priority over the replacement of these high-TOC subsystems. To put the issue in perspective, the shortage of funds available to address the aging avionics problem is so acute that an option under consideration by Air Combat Command is the early retirement of the F-117 stealth fighter because of insufficient funds to replace the infrared acquisition and designation system (IRADS), the color multipurpose display system (CMDS), and the electronic data transfer system (EDTS), all of which are facing obsolescence problems.

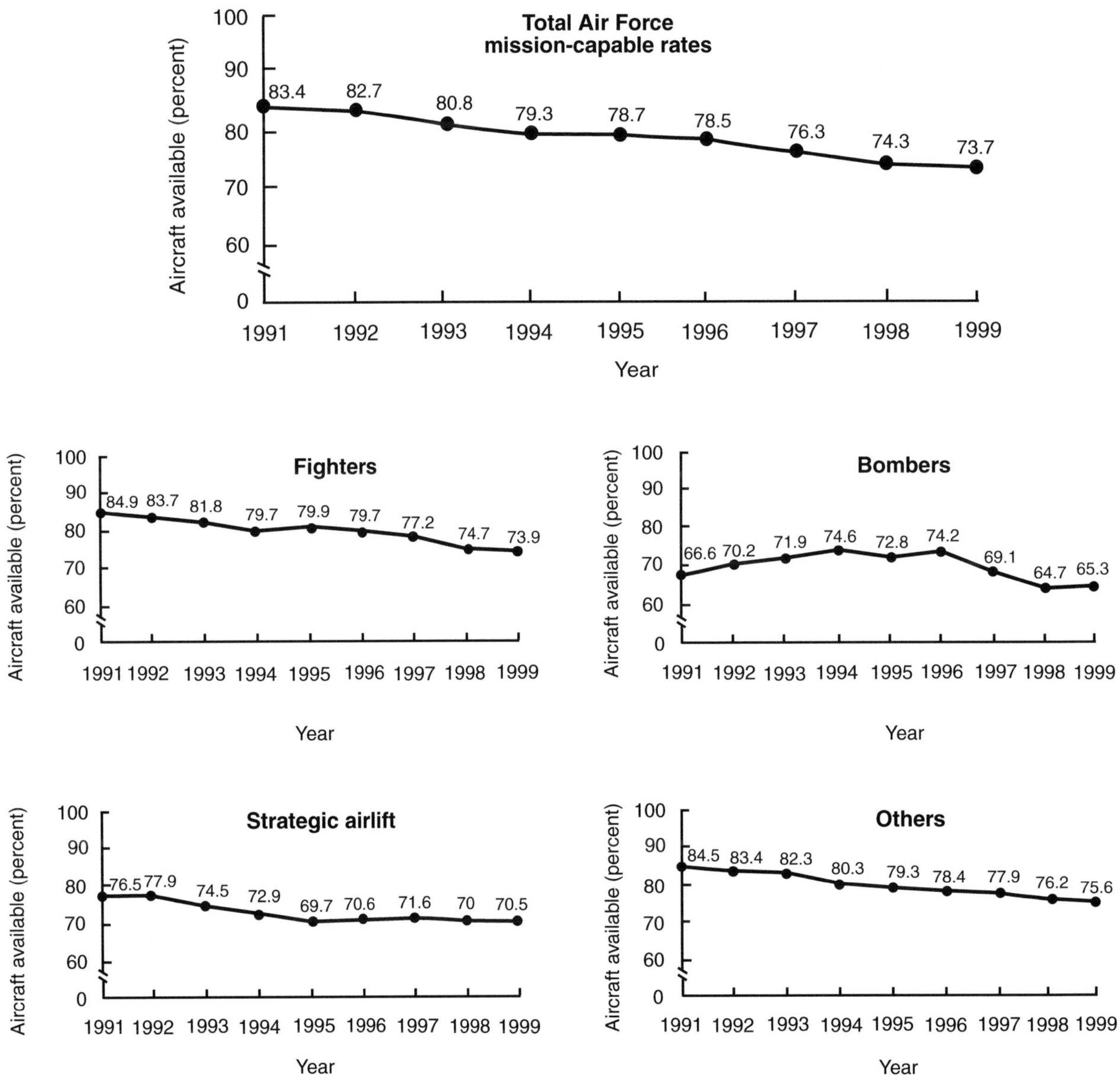

FIGURE 2-5 Declining Air Force mission-capable rate. Source: Air Force Magazine, 1999.

TABLE 2-2 Aircraft Currently in Service

Type	Quantity	Role
USAF		
A-10 Thunderbolt II	127	Close air support/ forward air control
OA-10 Thunderbolt II	99	Close air support/ forward air control
B-1B Lancer	81	Strategic bomber
B-2A Spirit	19	Strategic bomber
B-52H Stratofortress	85	Strategic bomber
C-5A Galaxy	28	Transport
C-5B Galaxy	50	Transport
C-5C Galaxy	2	Transport
C-9A/C Nightingale	23	Transport
C-12 Huron	36	Transport
C-17A Globemaster III	41	Tactical transport
C-21A (Learjet 35A)	79	Transport/ communications
C-23A Sherpa	3	Freight transport
VC-25A (Boeing 747)	2	Presidential transport
C-27A Spartan (G.222)	10	Transport
C-130E/H/J Hercules	183	Transport
EC-130E/H Hercules	22	Electronic intelligence
AC-130H/U Spectre	21	Gunship
MC-130E/H/P Hercules	66	Special operations
NC-130 Hercules	4	Test and evaluation
C-135A/B/C/E	7	Transport
EC-135	12	Electronic intelligence
OC-135	3	"Open Sky" Treaty
RC-135	20	Reconnaissance
KC-135	249	Tanker
NKC-135	2	Tanker
C-137B/C	6	VIP transport
C-141B Starlifter	95	Transport
E-3B/C Sentry	32	AWACS
E-4B (Boeing 747)	4	AACP
E-8C J-STARS	5	Surveillance
E-9A (DHC Dash-8)	2	Range surveillance
EC-18B/D (Boeing 707)	4	Reconnaissance/ surveillance
F-15A/B/C/D Eagle	404	Fighter
F-15E Eagle	201	Fighter/attack
F-15A/B/C/D Eagle	14	Test
F-16A Fighting Falcon	3	Fighter/attack
F-16B Fighting Falcon	18	Fighter/attack
F-16C Fighting Falcon	568	Fighter/attack
F-16D Fighting Falcon	88	Fighter/attack
F-117 Nighhawk	52	Attack
KC-10A Stratotanker	59	Tanker
RQ-1A Predator (UAV)	6	Reconnaissance/ surveillance
TG-3 (glider)	3	Reconnaissance/ surveillance
TG-4 (glider)	14	Reconnaissance/ surveillance
TG-7 (glider)	9	Training
TG-9 (glider)	4	Reconnaissance/ surveillance
TG-10 (glider)	1	Training
TG-11 (glider)	2	Training
HH-1H Iroquois	8	Missile support
UH-1N Iroquois	64	Missile support
TH-53A Sea Stallion	6	Search and rescue
MH-53J Sea Stallion	40	Special operations
HH-60G Black Hawk	54	Search and rescue
MH-60G Black Hawk	10	Special operations
T-1A Jayhawk	183	Training
T-3A Firefly	111	Training
T-37B Tweet	415	Training
T-38A Talon	414	Training
AT-38B Talon	78	Training
T-41 Mescalero	3	Training
T-43A (Boeing 737)	10	Training
CT-43A (Boeing 737)	1	Training
TC-18E (Boeing 707)	2	Training
UV-18 Twin Otter	2	Parachute training
U-2R/S	31	Reconnaissance
TU-2R/S	4	Training
WC-130H/W	3	Weather reconnaissance
Total	**4,307**	
Air Force Reserve (AFRES)		
A-10 Thunderbolt II	27	Close air support
OA-10 Thunderbolt II	27	Forward air control
B-52H Stratofortress	9	Strategic bomber
C-130 Hercules	112	Transport
C-141B Starlifter	44	Transport
C-5A Galaxy	32	Transport
F-16C Fighting Falcon	56	Fighter/attack
F-16D Fighting Falcon	8	Fighter/attack
HH-60G Black Hawk	21	Special operations/ search and rescue
KC-135E/R	75	Tanker
WC-130H/W	10	Weather reconnaissance
Total	**421**	
Air National Guard (ANG)		
A-10 Thunderbolt II	78	Close air support
OA-10 Thunderbolt II	18	Close air support
B-1B Lancer	14	Strategic bomber
C-5A Galaxy	14	Transport
C-130 Hercules	215	Transport
C-141B Starlifter	16	Transport
C-21 (Learjet 35A)	4	Transport/commun.
C-22B (Boeing 727)	3	Transport
C-26A/B (Metro III)	11	Transport
F-15A/B/C/D Eagle	90	Interception
F-16A Fighting Falcon	102	Fighter/attack
F-16B Fighting Falcon	26	Fighter/attack
F-16C Fighting Falcon	340	Fighter/attack
F-16D Fighting Falcon	29	Fighter/attack
HH-60G Black Hawk	17	Special operations/ search and rescue
KC-135 Stratotanker	225	Tanker
T-43 (Boeing 737)	2	Training
Total	**1,204**	
Grand Total	**5,932**	

Source: U.S. Air Force, 2000b.

DECLINING READINESS

Beyond the problem of rising maintenance/logistics costs and insufficient resources for modernization is the fundamental issue of combat and mobility readiness.

> The Air Force reports that mission-capable rates for its aircraft have declined by 10 percentage points—from 83 percent to 73 percent—since 1991. And rates of cannibalization (a measure of how often maintenance crews must take a part off one aircraft to maintain another) increased by 78 percent between 1995 and 1998, indicating a shortage of spare parts (CBO, 2000).

These data are illustrated in Figure 2-5. Although the committee does not have specific data linking the decline in readiness to aging avionics, the fact that avionics maintenance accounts for approximately one-third of total aircraft maintenance costs supports this conclusion. Air Force officials from the Air Combat Command and Air Mobility Command interviewed by committee members confirmed the linkage (personal communications with Brig. Gen. Randolph Bigum, director of requirements, Air Combat Command; and Maj. Gen. Michael Wooley, commander, Tanker Air-Lift Control Center, Air Mobility Command, September 26, 2000).

The magnitude of the Air Force's aging avionics problem cannot be fully comprehended without considering the diversity of types of aircraft flown (68 in the Air Force, 11 in the Air Force Reserve, and 17 in the Air National Guard), the small fleets of some types of aircraft (e.g., only 1 CT-43A), the multiple versions of the same aircraft (e.g., F-15 A, B, C, D, and E), and multiple users of the same aircraft (e.g., A-10 used by Air Force, Air Force Reserve, and Air National Guard). In light of these data (Table 2.2), the committee concluded that the magnitude of the aging avionics problem is large and is growing. This urgent problem must be addressed by Air Force management through enterprise management supported by informed program management.

3

Current Activities and Programs

During the course of the study, the committee reviewed documents and attended presentations by representatives of government agencies and industry on the key issues associated with aging avionics. Numerous government and industry organizations are addressing these issues from several different perspectives: policy making; process initiatives; open systems development; legacy system upgrades; DMS/OP; development of tools; and the reengineering or remanufacturing of obsolete parts. Policy changes relate to organizational structures, budget processes, laws, and other factors. Some initiatives have been established to address processes for replacing obsolete avionics (or avionics that will soon be obsolete) either through technology refresh of components or boards or through equipment upgrades. Still others are developing tools to track DMS/OP and reengineering methods so that systems can be maintained after the loss of a manufacturing source. Most organizations are searching for technology changes and advances to reduce or avoid DMS/OP, through so-called open systems, through legacy system upgrades that do not require major system replacements (e.g., software wrappers), or through technologies to remanufacture parts. Most of these activities are being done independently with little or no coordination. As an old saying goes, "a thousand flowers are blooming, each in itself a thing of beauty, but there is no plan or design for the garden."

Table 3-1 summarizes many activities and programs attempting to resolve the DMS/OP problem (see Appendix A for brief descriptions). No doubt, there are numerous other activities of which the committee is unaware. As the table shows, many government organizations have adopted similar approaches or are funding similar projects, some of which are duplicative and/or overlap. Although there is a good deal of activity in the area of process initiatives, economies of scale are being missed because of the lack of coordination. The successful obsolescence-management program at the air logistics center for the F-15 at Warner Robins Air Force Base, is a case in point. This program, which is recognized as a model program by the Air Force, uses the Air Force Research Laboratory (AFRL) Hub (a centralized communications network) and a commercial tool called avionics component obsolescence management (AVCOM) supplied by Manufacturing Technology, Incorporated. The B-2 program, another recognized leader in obsolescence management, uses a different commercial tool supplied by TACTech and does not participate in the AFRL Hub. These two programs are following their own courses ("stovepiping"), which prevents them from sharing information and finding common solutions to common problems.

Another example of stovepiping is the Government-Industry Data Exchange Program (GIDEP), chartered to be DoD's centralized database for managing and

TABLE 3-1 Current DMS/OP Activities

				Technology				
Organization	Activity	Policy Making	Process Initiative	Open Systems	Legacy Upgrade	DMS/ OP	Tools	Reengineering
OSD	Open Systems Joint Task Force (OS-JTF)	X	X	X				
	Joint Technical Architecture (JTA)	X	X					
	Joint Strike Fighter		X	X				
	Defense Advanced Research Projects Agency (DARPA) Information Technology Office (ITO)		X	X	X			
	Defense Microelectronics Activity	X	X			X	X	X
	Diminishing Manufacturing Sources Material Shortages (DMSMS) Teaming Group		X			X	X	X
	Government Industry Data Exchange Program		X			X		
	Defense Logistics Agency (DLA) Generalized Emulation of Microcircuits					X	X	X
	DLA Shared Data Warehouse		X			X	X	
USAF	Aeronautical Systems Center Affordable Combat Avionics Office	X	X	X	X			
	Air Force Research Laboratory (AFRL) DMSMS Program	X	X			X	X	
	AFRL Manufacturing Technology Electronics Parts Obsolescence Initiative	X	X		X	X	X	X
USN	Naval Aviation Systems Team		X	X	X			

USN	Navy Helicopter Modernization Program			X	X			X
	Diminishing Manufacturing Technology Center	X	X			X	X	
USA	Aviation Electronic Combat Office			X	X	X		
	Modernization Through Spares/Continuous Technology-Refresh Program		X		X	X		
	Aviation Applied-Technology Directorate	X	X	X				
	Rotary Open-System Architecture			X				
Industry	Lockheed Martin Proven Path and Systems, Technologies, Architecture, and Acquisition Reform Study		X	X	X	X	X	
	Boeing Bold Stroke and Open Avionics Systems Integration Study		X	X	X	X	X	
	Avionics-Component Obsolescence Management					X	X	
	Transition Analysis of Component Technology					X	X	
	SMART Part		X	X	X	X		X
	National Rotorcraft Technology Center	X	X		X	X		
	Software Engineering Institute		X	X			X	
	National Center for Advanced Technologies	X	X					
	University of Maryland Computer-Aided Life-Cycle Engineering (CALCE) Center		X				X	

sharing information on DMS/OP material shortages among DoD and industry groups to reduce redundancies and improve effectiveness. Today, however, Air Force, Army, and Navy organizations, as well as commercial companies, provide their own alerts, health analyses, and recommended solutions. GIDEP has never been fully used. Capable manufacturers feel that the visibility of component issues (DMS) across multiple products and platforms would present a real opportunity to create innovative and cost effective solutions to parts and service problems.

In the area of open systems architecture, the committee noted that the Software Engineering Institute (SEI) (a federally funded R&D center sponsored by DoD through the Office of the Under Secretary of Defense for Acquisition, Technology, and Logistics) has been actively involved in work on open systems since 1993, developing tools and process initiatives, and developing formal standards. However, neither DoD nor the Air Force has taken full advantage of advances in software architectures. The committee identified more than 25 organizations, both within and outside the Air Force, that are working on various aspects of the DMS/OP problem. Although each organization may be effective in its limited area, overall coordination of these activities is loose, at best, and the results are not broadly distributed to DoD or the Air Force. With a coherent DoD/Air Force strategy for dealing with DMS/OP problems, a collective management could be established for these diverse activities, which could lead to more productive use of the results and minimize redundant expenditures.

4

Analysis of the Current Situation

The previous chapters have described the nature and extent of the aging avionics problem, as well as ongoing efforts to address it. The problem, which significantly affects the readiness of Air Force assets, is manifested most severely in the maintenance and modernization of the current fleet. However, the obsolescence of avionics systems in new aircraft will also have to be managed. New approaches to avionics design, such as MOSA, can mitigate the aging problem in future aircraft, and MOSA precepts can also be incorporated into programs to modify and upgrade legacy avionics systems—especially if the added short-term expense is justified by long-term savings in TOC. However, the path to better management of the aging avionics problem is strewn with many obstacles, of which the technical obstacles may be the easiest to overcome. A comprehensive solution will require that more challenging obstacles, such as fragmented management responsibilities within the Air Force, budget/funding restrictions in aircraft modernization programs, and ensuring the maintenance of a healthy and dynamic industrial base of avionics suppliers, will also have to be addressed.

This chapter presents the committee's analyses of major challenges to solving the problem of aging avionics in the four most important issue areas: Air Force enterprise-management processes; budget/funding management processes; technical issues; and business issues.

GOVERNMENT ENTERPRISE-MANAGEMENT PROCESSES

The coordinated management of the aging avionics problem in a complex organization like the Air Force and a rapidly changing technological environment raises several concerns: fragmented responsibilities in DoD for managing the problem and activities that cut across all of the services; attracting and retaining support personnel for new and upgraded weapon systems; training air crews and support personnel for new systems; and keeping track—and control—of constantly changing avionics configurations in the fleet.

Fragmented Responsibilities

The committee determined that no enterprise (DoD-wide or even Air Force-wide) strategy has been developed for dealing with the aging/obsolescent avionics problem. The committee found little evidence of effective cross-program, cross-platform, or cross-service coordination in the planning, funding, or implementing of actions to maintain the existing fleet or to define/design maintainable, extensible avionics systems for the future. In fact, the lack of coordinated management is evident even in activities dealing with multiple versions of the same platform.

Evolution of the Air Force Weapon-System Management Structure

Fragmented management is the result of recent reorganizations within the Air Force. The Goldwater Nichols Act of 1986 (P.L. 99-433) required that all Acquisition Category I (ACAT I) programs[1] be assigned to program managers who report directly to a program executive officer, who reports in turn directly to the service acquisition executive. ACAT I programs are considered major defense acquisition programs (MDAPs) and generally entail total expenditures of more than $365 million on R&D, testing, and evaluation in FY00 constant dollars or more than $2.19 billion for procurement. Managers of lower cost acquisition programs (ACAT II, III, or IV), including acquisition programs for most avionics systems, report to a product center commander or an air logistics commander, both of whom report to the commander of Air Force Materiel Command (AFMC). Although one way to reduce TOC would be to establish a common avionics system for multiple aircraft, the program manager of an avionics system and the program manager of an MDAP (who controls very large procurement funds) are each concerned about management of their own programs (stove-pipe management) and have different reporting chains of command.

At the same time the Air Force was reorganized to comply with the Goldwater Nichols Act, it was also in the process of combining the Air Force Systems Command and the Air Force Logistics Command into the AFMC. In response to concerns that management of programs for aircraft acquisition and support would become too fragmented, the AFMC created the concept of integrated weapon-system management (IWSM) to coordinate the acquisition and support of all Air Force programs in the AFMC. Although IWSM provides an effective approach to coordinating management for the total life cycle of a single weapon system or aircraft platform, it does not have a mechanism for addressing problems that affect multiple aircraft platforms. The management structure of operational logistics for fleet avionics is similar. Therefore, maintenance is largely reactive to crises. The current management structure does serve the basic purpose of providing integrated management for each weapon system. However, stronger horizontal management authority for issues like aging/obsolescent avionics will require some form of matrix management. Integrated product and process teams or special program offices are examples of management techniques that could be used.

[1]Aircraft platform programs, such as a fighter, bomber, or transport aircraft, are categorized as ACAT I because of their large total acquisition cost.

Importance of Avionics Modernization Road Maps

A practical, affordable approach to assessing and managing the problem in terms of a single platform must begin with the preparation of avionics modernization road maps for each platform, emphasizing planned, periodic upgrades. If supported by a concurrent budgeting plan, a series of cost-effective, systematic, periodic (every two to three years) upgrades could then be planned to upgrade system performance incrementally and/or to mitigate future obsolescence problems and ensure a ready, highly capable fleet.

The Air Force is now (and should continue) creating and implementing road maps for each platform (Raggio, 2000). The test of success of these road maps will be how well they are coordinated with the budgeting process—especially out-year commitments. These road maps must become real plans, rather than "wish lists" that never make it above the funding cut line.

Comprehensive road maps for individual platforms can also provide an effective framework for cross-platform and, eventually, cross-service coordination. The road maps could also be used as a basis for total-enterprise planning and management, which could reduce redundant expenditures and improve schedule efficiencies. The sharing of best practices among different weapon-system programs would be an added benefit. In fact, from a process viewpoint, the coordinated management of the activities identified in Chapter 3 of this report could reduce redundancies and increase efficiencies for all current activities addressing the aging avionics issue.

Coordinating MOSA Management with the Joint Technical Architecture

The platform-to-platform interoperability requirements for new and legacy weapon systems are imposed by the Joint Technical Architecture (JTA) Development Group and the Global Information Grid (GIG).

The development of the JTA is implemented under a council managed jointly by the Office of the Assistant Secretary of Defense (Command, Control, Communications and Intelligence (OASD[C3I]), the Office of the Under Secretary of Defense for Acquisition, Technology and Logistics (OUSD[AT&L]), and the Office of the Joint Chiefs of Staff (OJCS). The GIG is managed by the Chief Information Office of DoD, currently headed by ASD (C3I).

However, the JTA Development Group was established to address *interplatform* interoperability issues. The latest version of the JTA (3.0) designates architectural standards that extend beyond the interoperability domains of weapons systems—imposing standards that apply to *intra*platform architectures (DoD, 1999a).This extension went beyond the JTA's initial charter (DoD, 1999b). This leads to the potential for generating inconsistent and/or conflicting approaches to the modular open-system design of avionics systems that is being guided by the Open Systems Joint Task Force (Logan, 2000).

The committee is quite concerned that the 359 mandatory standards being imposed on avionics systems by JTA (along with another 228 standards being evaluated for mandatory designation) could be counterproductive because they could inhibit design innovation and minimize the achievement of MOSA goals. In addition, they would have little or no real effect on platform-to-platform interoperability.

In the committee's opinion, JTA requirements should be limited to *inter*platform interoperability standards and should not include *intra*platform architectures and standards. In addition, tightly coordinated management for DoD architectural standards is necessary, rather than the loosely integrated leadership of JTA, coupled with the somewhat disconnected management of the JTA, GIG, and MOSA.

Need for New, Innovative Contracting Approaches

Current DoD contracting approaches generally do not provide positive incentives to either government contracting offices or defense contractors for reducing product cost *to the government*. Production programs frequently use firm fixed price contracts, which do put a cost cap on the government's obligation; the contractor benefits from any savings (or pays for overruns), but there is no reduction in cost to the government. Any postcontract change in requirements generally results in as large a bill as the contractor can justify. On subsequent contracts, the government strives to reduce the price based on previous actual costs, and the contractor attempts to justify the highest price traffic will bear. This often leads to adversarial relationships that undermine joint attempts to reduce costs. The recent use of award fees is a step in the right direction, but the most successful approaches are based on positive incentives through shared savings (Ebersole, 2000).

Contracts that involve generous shared savings of any cost *to the government* would create a "win-win" environment for all participants. With the advent of MOSA, cost reductions will be more likely, and both government and industry will be motivated to seek innovative ways to improve their performance. A mutually agreed upon TOC model can also provide incentives for reductions in TOC.

Management Focal Point

The committee received extensive briefings on the aging avionics problem from the Aeronautical Systems Center headed by Lt. Gen. Robert Raggio of AFMC. With its Affordable Avionics Initiative, the Aeronautical Systems Center has already become a strong focal point in the Air Force for addressing the aging avionics problem. Recently, the Affordable Avionics Initiative was placed under the authority of the newly created Aging Aircraft System Program Office, which will be led by a general officer and will include all aspects of aging, with an emphasis on aging avionics. This new office could become the starting point for the development of an Air Force-wide enterprise strategy for addressing the aging avionics problem. However, this office will need significantly more funding to halt the upward spiral of avionics support costs.

Each service has its own management processes for making budgetary decisions and overseeing programs. In the past, the Air Force used Quarterly Acquisition Program Reviews (QAPRs) to evaluate the potential of modular open avionics designs to reduce avionics TOC. The committee believes the Air Force should continue to use QAPRs as a tool for periodic top-level Air Force oversight and management of the problem of aging avionics.

Education and Retention of Qualified Personnel

As advances in computer technology have been incorporated into modern avionics systems, the software content in these systems has increased dramatically;

thus, more software engineers must be hired and trained to accommodate this trend. Recent studies have shown that the technical competency of maintenance personnel is eroding, particularly in the software technology area, in both the government and the defense industry (NRC, 1997). The maintenance workforce is aging and falling further and further behind the continuing technology surge. Workforce turnover is a related problem, as younger, high-potential personnel are leaving for jobs in nondefense industries where pay scales are higher and perceived opportunities are more abundant.

The committee was in general agreement that this software environment will continue to degrade in the future. The growing disparity between the supply and demand of software engineers, the mismatch between old and new design methodologies faced by an aging workforce, the tug of war between government and industry for skilled personnel, and the increasing complexity of software system, will create an even more serious problem for avionics maintenance in the future.

Education and training of software-proficient system engineers, along with an efficient design environment (e.g., tools and processes) is, and will be, essential. The Air Force should consider ways to identify and share best practices in software design/maintenance and explore ways to consolidate software support activities that will result in a critical mass of technical talent. To attract new technical personnel, the Air Force should expand its hiring incentives to potential employees that could help narrow the gap between government and industry opportunities.

The committee members had different opinions about joint actions by the Air Force and industry to deal with this problem. Much of the debate centered around the degree to which the Air Force should contract out to industry a substantial portion of avionics software maintenance, especially for legacy systems currently supported by in-house (organic) capability. In some cases, only government depots have the detailed data and experienced personnel familiar with legacy equipment, making it difficult for industry to assume maintenance responsibility. However, in industry, personnel are increasingly being used by government depots as on-site workers to compensate for the diminishing in-house capability.

Because most of the major upgrade and new avionics software (in a MOSA environment) will be designed by avionics suppliers and platform prime contractors, both industry and government will have to work together adjusting continually to achieve a balance in the future. Therefore, the committee decided not to offer an all-encompassing recommendation regarding software outsourcing at this time. However, this issue deserves continuing attention.

Training

The effects of aging avionics on the training of Air Force personnel are not currently being measured, managed, or included in aging avionics models. Relevant activities include: training logisticians to develop and execute plans across platforms to solve aging avionics problems; training configuration management specialists to keep track of all changes and their downstream effects; and training mission planners to consider the effects of changes in avionics on sortie rate and mission performance.

The acquisition of systems for all of the active and reserve services is performed by two groups of personnel: (1) program managers and engineers in system program offices (SPOs) for systems under development; (2) item managers at depots for systems in production or operation. For MOSA to be effective for the acquisition of upgraded or new systems, both groups must be trained in defining MOSA requirements, writing subsequent specifications for systems, evaluating bids, and developing and administering contracts.

SPO personnel receive their training at the Defense Systems Management College or the Industrial College of the Armed Forces. Curricula at both institutions should be reviewed to ensure that courses are consistent with MOSA. Depot personnel at the Air Force's three logistics centers (Oklahoma City, Ogden, and Warner Robins) may attend either institution and may receive additional in-house training on special topics. Depot personnel involved in acquiring, developing, defining, or specifying software requirements should receive in-house training in MOSA.

BUDGETARY ISSUES

The AFMC estimates that $250 million to $275 million per year in additional funding will be necessary to address the aging avionics issue (personal communication with Lt. Gen. Robert Raggio, Commander, Aeronautical Systems Center, October 6, 2000). By any measure, this is a large amount of money; however, even with this amount, government constraints on allocations and expenditures could preclude its being spent in the most efficient way. The committee noted four budgetary issues that need attention:

- Long acquisition and upgrade cycles virtually require that avionics technology-refresh cycles be built into program plans during the engineering, manufacturing, and development phases prior to initial fielding.
- Budgets are compartmentalized into rigid accounts ("colors of money"), which makes it difficult for managers to address problems as they arise.
- Many system-capability upgrades could be opportunities to solve avionics obsolescence issues if a combined approach were used. However, certain mandated flight-safety upgrades have little impact on solving the problems caused by high-TOC avionics subsystems.
- Obtaining front-end funding to reduce TOC is difficult.

Long Acquisition and Upgrade Cycles

Avionics technology is advancing at a much faster pace than DoD acquisition cycles (Figure 4-1) because avionics product cycles are driven by the commercial market, whereas DoD acquisition cycles are complex and often delayed by funding constraints (DSB, 1999). Perhaps the best example is the F-22, which will have undergone four avionics technology-refresh cycles before the first production airplane rolls off the line. Another example is the F-15 APG-63 radar modification. The contract was awarded in FY97, and the first unit was delivered in FY99. Because of funding constraints, production will cease in FY04 and FY05 and resume in FY06 (Donatelli, 2000). As a result, unless funds are reprogrammed in future budgets, the modification will not be fully installed until FY09. Unless the manufacturer is funded to procure all parts during the initial years of production, the interval of 11 years from first to final installation almost guarantees that the parts will be obsolete in future years. Unfortunately, MOSA was not incorporated into this redesign, so the cost of changes will be higher than they might have been.

In areas like avionics in which technologies are evolving rapidly, it makes little sense for design and implementation cycles to stretch out for many years. New designs or retrofitting modifications must be planned and implemented when relevant technologies are available. Therefore, at all levels of the system, parts/functional updates must be planned that minimize impacts on the unchanged hardware/software. In most cases, the window of availability is about five years, or at most ten years. To complete avionics modification

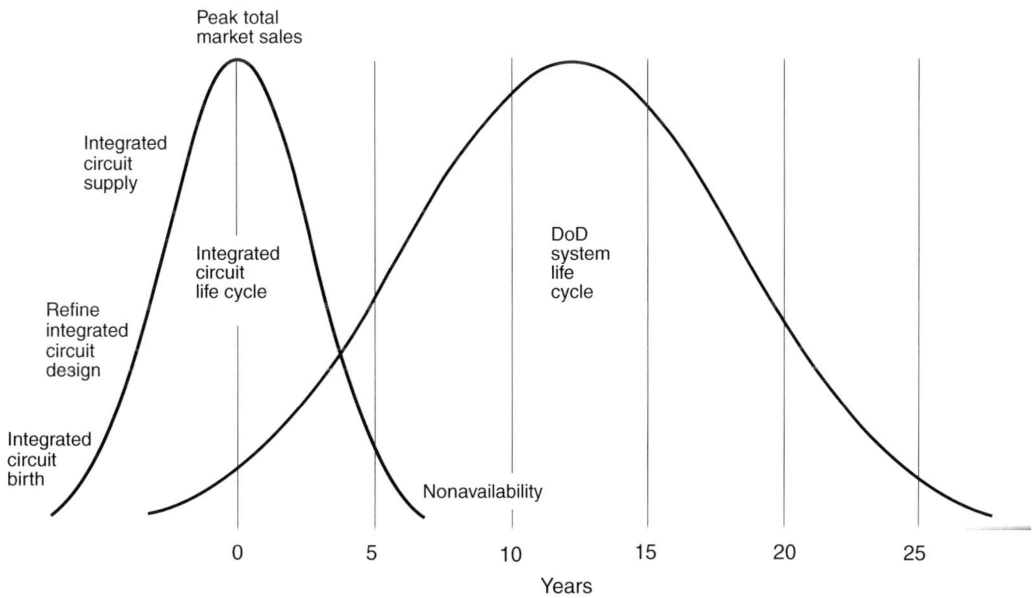

FIGURE 4-1 Life-cycle mismatch. Source: Wasson, 2000.

programs within this time window, DoD will have to change the current planning and implementation cycle in several ways.

In Directive 5000.1, DoD defined principles for acquisition reform to reduce cycle times (DoD, 2000b). These "evolutionary acquisition" principles are based on a recognition that firm requirements for the entire life cycle of a weapon system cannot be specified realistically from the beginning; instead, the requirements must be developed interactively, based on the experience of warfighters using the system. Under this doctrine, delivery of the first unit would not be delayed until all mission requirements were met; the first unit would be a functional, but basic, system that would be upgraded at regular intervals through "block upgrades" of hardware and software until all requirements were satisfied. Each block upgrade would provide an opportunity to capture the most up-to-date technology. Periodic upgrades, which would be specified in evolving platform road maps, could be an effective strategy for addressing the aging avionics problem, especially if it were combined with MOSA architectures that emphasize ease of change.

Colors of Money

To ensure that appropriated funds are used for their intended purpose, Congress has placed a number of legal restrictions on funds available to address the aging avionics problem:

- Project requirements of a specific fiscal year must be funded only with appropriations enacted for obligation in that fiscal year.
- The purpose of the expenditure must be authorized in the appropriation.
- Amounts appropriated for general or specific purposes may not be exceeded even if changing priorities dictate otherwise.

The major budget categories associated with aging avionics are: research, development, testing, and evaluation (RDT&E, designated as 3600 funds); procurement (designated 3010, 3011, 3020, or 3080 funds); and O&M (designated 3400 funds). Although in practice the lines between these categories are somewhat blurred, managers cannot use funds appropriated in one account to solve a problem associated with another account. For example, even though funds may be available to procure an avionics system in a given year, they cannot be used to solve a lingering RDT&E problem with that system. Commercial corporations make such decisions on a routine basis for the benefit of the enterprise, but DoD program managers are denied this flexibility. As a result, program managers spend a great deal of time trying to manage these "color of money" issues. In some cases (e.g., the fiscally constrained F-15 program), the amount of money in the various budget categories actually determines what can be done toward meeting avionics requirements, rather than the reverse (Durante, 2000).

In each budget category, there are many competing demands, some of which take priority over reducing the TOC of avionics systems. For example, many of the avionics modifications that are funded, such as the installation of TCAS and the reduced vertical separation minimum (RVSM), are capability improvements required by the Federal Aviation Administration (FAA) for all aircraft that fly in the new global air traffic control system. Few of these modifications will affect avionics with high TOCs. The Air Force needs a systematic funding that addresses both aging avionics components and capability improvements.

All of the funds available for avionics modernization combined still leave a shortfall that will continue to increase unless the budget is increased to provide funds to support the flying-hour program and to meet the following needs:

- safety-mandated avionics upgrades
- avionics upgrades necessary to fly in the air traffic control system
- replacements of high-TOC avionics with new avionics, with the goal of achieving economic, consistent solutions throughout the fleet
- periodic, systematic upgrades that address the issues of performance and obsolescence

Front-End Funding

From the point of view of a program manager concerned with allocating the current year's budget, the least expensive approach to fixing an avionics problem is a customized, point solution for that problem. The likelihood that this short-term solution will be difficult to maintain or upgrade and will, therefore, cost more over the life of the aircraft is not an immediate concern. A more comprehensive solution, such as the application of MOSA to replace obsolete hardware and software, generally requires more expensive redesigning and requalification. In many instances, this solution will cost more in the short run but will reduce TOC in the long run. Unfortunately, in the current budget environment, it is difficult for the program manager to justify spending money now to save money later, especially when savings may occur in a different budget category.

Since 1995, the Air Force has kept track of high-TOC avionics components in its AFTOC database (U.S. Air Force, 2000a), and since 1998 a small amount of funding has been made available for the front-end funding of projects designed to reduce TOC. The committee supports these efforts, but additional funding should be made available to support well documented opportunities to reduce avionics TOC. Revolving funds, such as the Working Capital Fund, to which all of the services contribute, may provide a possible mechanism for providing these funds.

Funding for initiatives to solve the issues related to DMS and to reduce TOC will remain a challenge for all of the services. The best solution would be to budget for and fund modernization modifications that incorporate modular open-systems architectures. This would not only solve immediate DMS problems but would also reduce the cost of solving future DMS problems.

The second-best solution would be for each service to create a budget line item that would be funded each year and used to solve DMS problems as they are identified. This line item would also fund other TOC cost-reduction initiatives. A rigorous cost analysis comparing the TOC for continued system operation with and without the change or initiative would have to be done to determine the comparative cost/benefit of competing proposals or initiatives.

The next best way to ensure that funds are available for solving DMS issues would be to use a revolving fund, such as the Working Capital Fund. However, the SPO director, the Air Logistics Command supply-chain manager for the commodity group, and the Major Command would all have to agree on the need for funding and for the surcharge increase the weapon system would incur by using the Working Capital Fund. The Air Force has already established a parts-obsolescence funding line as part of the Working Capital Fund overhead. However, this line item, which was only established in FY00, is relatively small.

TECHNICAL ISSUES

The concept of MOSA for avionics design has evolved in DoD and industry over the past several years. Generally patterned after the architecture of modern commercial information systems, the purpose of MOSA is to provide scalable, extendable, modular avionics systems that can be upgraded affordably by the replacement of modules.

Figure 4-2 shows the hierarchical structure associated with the Joint Strike Fighter; levels three through six represent the avionics suite. Various levels of modules, or building blocks, are shown in a hierarchy, with defined functional/electrical/physical interfaces at the horizontal and vertical "flanges" where these units interconnect.

Military avionics systems have had traditionally "federated" structures: that is, assemblages of largely independent, single-function subsystems ("black boxes") that collectively met the overall performance requirements. Internal interfaces in a subsystem were owned and controlled by the designer/supplier (a

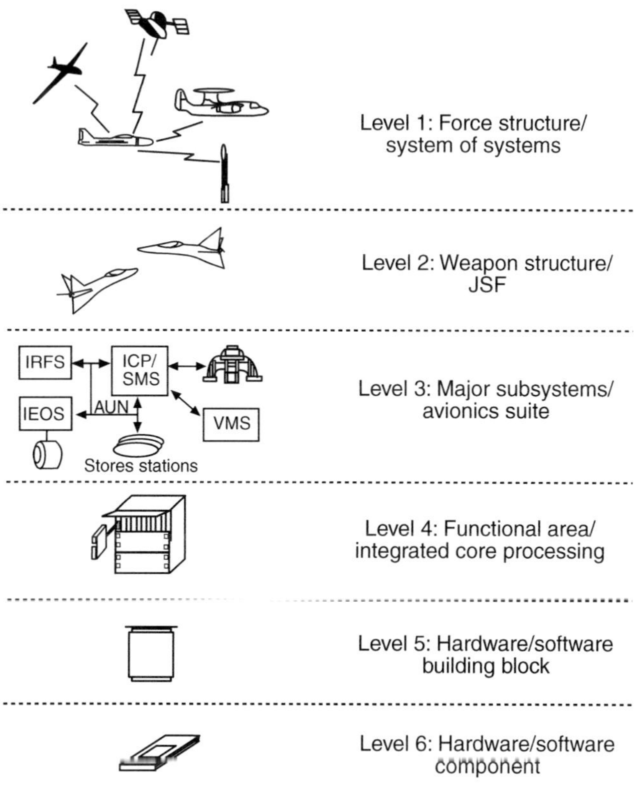

FIGURE 4-2 JSF architectural hierarchy. Source: Logan, 2000.

"closed" system); external interfaces (e.g., communication interfaces) were specified by the government.

As avionics have become more digitized, and as affordable computing power has increased, the trend has been to develop integrated avionics systems. In these systems, more functionality is embodied in software that could be integrated into one or more general-purpose processors. Hardware and software from multiple suppliers operate in an integrated environment. Thus, many internal interfaces have to be shared by the suppliers (less "closed" more "open"). A modular approach to the architecture would simplify the definition and control of these interfaces, leading DoD and industry to define and develop a MOSA environment for avionics system design.

Modular systems are commonly defined in terms of the portability of applications software and the implementation of widely supported standards for key system interfaces. In avionics architectures, technology independence is important, but modular partitioning is even more important. The primary goal is for systems to be affordable, as well as easily upgradable, expandable, and scalable.

Common Understanding of MOSA

To meet evolving military requirements for better sensors, countering an increasing variety of threats, and increasing mission capability, future avionics systems will have to be flexible and easy to modify and upgrade. The Air Force has endorsed a MOSA approach as a strategy for managing the problem of avionics obsolescence in both new systems and legacy aircraft systems. Office of the Under Secretary of Defense for Acquisition, Technology and Logistics has chartered the Open Systems Joint Task Force to coordinate and motivate DoD MOSA activities and to solicit inputs from industry. The task force has characterized an open system as "a design based on nonproprietary interface standards broadly accepted and used throughout industry" (Logan, 2000). The architectural framework adopted by the task force is shown in Figure 4-3.

A more definitive characterization of an "open system" is described in *Architectures for Next Generation Military Avionics Systems* (Borky et al., 1998). Open systems are also generally modular, but with additional attributes. A modular system has the following attributes:

- The system is designed to maintain external hardware/software interface compatibility of a module independent of changes made internal to the module.
- Both hardware and software (physical and functional/logical) aspects of architectural interfaces are included in the system.
- The system can be scaled in capability by incrementally adding or deleting modules of functionality.

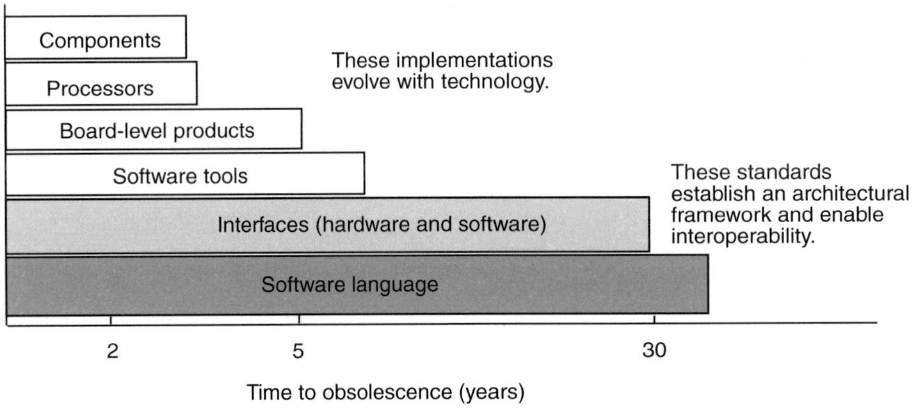

FIGURE 4-3 Architectural framework adopted by the Open System Joint Task Force. Source: Logan, 2000.

- The system can be maintained or upgraded by selective replacement of elements without impacting other elements.
- The system can reuse existing elements and provide reusable elements to other systems.

An open system has the following attributes:

- All of the attributes of a modular system are also included in an open system.
- The system can be integrated from elements supplied by multiple sources.
- Choice/application of standards represent a design decision that follows open system partitioning and functional interface definition.

Subsequent to modular/open architecture definition, the requirements for interoperability, communications protocols, data formats, and logical functionality can lead to the definition and selection of proper open standards.

The committee generally agrees that a transition to MOSA for avionics system designs would significantly improve the aging/obsolescence/upgrade environment. However, perceptions of MOSA vary considerably throughout DoD and industry, especially concerning the definition of "open." Definitions range from truly open architectures with standard and/or publicly available interfaces at most hardware and software levels of the structure to architectures with proprietary ("locally open") interfaces at the modular (line-replacement unit/building block) level and open interfaces at the higher functional, subsystem, and intersystem levels.

Although the words "modular" and "open" are almost always used together—as in the acronym MOSA—the committee wishes to stress that they are, in fact, separate concepts. Although open systems would in principle provide great benefits to the Air Force, the associated business complications could make their wide-scale application problematic. Purely "modular" but "closed" or only partially open systems (i.e., systems available from only one vendor) in most cases can provide most of the benefits of MOSA without the complications and thus may be more easily applied in a variety of situations.

Complicating the issue is that—for various reasons (e.g., technology, cost, size/weight, legacy systems)—avionics architectures will continue to be a mix of federated systems ("black box" functionality) and integrated systems (common processing and/or sensors for multiple functions) for the foreseeable future. The degree of hardware/software independence/decoupling will vary, and the disciplines for defining subsystems, modules, and components will also vary. In all likelihood, the perception of openness will continue to be fuzzy. Therefore, the degree of openness will necessarily vary from system to system, and the emphasis on mitigating the problems of aging/obsolescent avionics and planned avionics upgrades should be on the *modular* (rather than the *open*) aspect of avionics architectures.

The modular approach should be extended to include nondigital functions, such as sensors, communication, and navigation systems which often dominate the cost of an avionics system. Openness should be required at the intersystem level to ensure interoperability and should be specified at interfunctional communications interfaces for which standard protocols provide the best technical and cost approach.

Although the MOSA concept will generally be used in designing new avionics systems, it can also be applied to the modernization of older, legacy systems. Obsolete and/or underperforming subsystems can and should (where practical and affordable) be replaced with newer, modularly structured subsystems. In addition to meeting current needs, MOSA, especially when compounded or repeated, could reduce the cost, time, and complexity of upgrades. Considering the long life cycle of a platform, and the short life cycle of avionics/electronics, future savings could be significant.

MOSA Design Tools

The committee found a critical need for further development and ensured availability of MOSA design tools that will support the disciplined design process for implementing avionics system architectures. These design tools include modeling and simulation tools and selected, high-level design languages (hardware and software) with related compilers.

A crucial enabler for the design and implementation of MOSA-driven avionics architectures is a set of "building codes" that govern system interfaces, module/functional definition processes, software languages, interconnect characteristics, and other global system parameters. Rigorous definition of modules, preferably through executable simulation objects, will be particularly important. The concept of building codes is known in software engineering as the use of "architectural patterns," which originate from the architectural domain (Alexander, 1977). A pattern is defined as an abstraction from a concrete form that can recur in a specific context. A pattern constitutes a template for problem-solving documentation, that describes the proven solution in a given context or problem. Rigorous definitions of modules representing established solutions to specific tasks necessary. To ensure the safety and reliability of avionics systems, formal verification methods and executable models will have to be used. According to a white paper, *Architectures for Next Generation Military Avionics Systems* (Borky et al., 1998):

> Object oriented design has emerged as the current leading methodology for managing complexity and ensuring correctness in system development. It seeks to improve on traditional approaches to functional decomposition and structured design through, among other things, better encapsulation of functionality, strict definition of interactions among entities, and traceability of requirements up and down a system hierarchy. This may well prove to be a powerful technique for achieving the open system characteristics needed in next-generation avionics suites.

The standards profile of a system emerges from the system-engineering process based on the most effective implementation at a given point in time. An evolving (but limited) set of standard interfaces could be considered for the development of new avionics systems to ensure the interoperability of subtier suppliers' equipment. Examples include commercial standards (e.g., ARINC, etc.) that enable dual-use applications. For fully integrated applications, commercial hardware and software standards should be considered to ensure interoperability among system components.

The Air Force and other services must recognize that developing a new modular design for a single platform will be costly. However, if a modular approach were adopted for several platforms, the savings could be significant. The modular approach would also provide design flexibility for addressing unique platform requirements through a common software language. To maximize the savings, the whole spectrum of tools including design tools, system enablers, building codes, and system modeling and simulation, could be addressed jointly by DoD and industry. The Air Force has already taken a step in this direction by working with the SEI (see Appendix A) on a system design and a maturity model of system engineering capability.

Database for the Reuse of Designs

An integral part of the MOSA design strategy will be providing easy access to reusable design fragments, including examples from which pieces can be extracted and examples of applied interface standards illustrating how they can be applied to solve difficult problems (especially the incorporation of advanced technology that might easily be considered incompatible or in conflict with the standards).

Reuse libraries should be indexed/accessible so a competent user can easily find relevant cases. Whether organized by case-based reasoning or some other method, they must engage a user in a dialogue that results in valuable suggestions that can be applied to real designs. Rather than a universal database accessible to all, reuse libraries should be developed by industry and should remain in their domain.

Configuration Management

Addressing DMS/OP and modernizing avionics systems in an era of constrained budgets raises a serious problem of configuration management. Because of reduced production buys and cyclical updates, aircraft fleets are not all in the same configuration. Program extensions often cause the same problem, especially at the component level where parts can become obsolete even before the retrofitting is complete.

Currently, the Air Force has no institutionalized configuration-management processes, tools, or requirements. In fact, configuration management is not even integrated with logistics, maintenance, training, testing, and operations processes. This issue will have to be addressed by a top-down management approach.

Streamlining Requalification/Recertification Testing

When avionics systems are modified or old components are replaced with new ones, the new system must be tested to prove its functional compatibility with existing hardware/software, to demonstrate the new performance, and to ensure flight safety. Testing represents an increasing percentage of costs for both DMS fixes and for system upgrades (e.g., certification testing has traditionally represented nearly 40 percent of software development costs). The lack of a practical, efficient (in time and cost) strategy for requalification and recertification testing of avionics systems and subsystems subsequent to modifications and/or upgrades is a major problem. The proper application of MOSA allows confinement of both failures and changes, forming the basis for the need for expensive, time-consuming regression testing after modifications or upgrades. The impact on TOC could be significant.

With advances in modern design tools, however, an affordable solution may be in sight. These tools provide high-fidelity emulation of detailed circuit operation, and, with some additional effort, could be formulated as stand-alone digital models of the product. Combined with the disciplined interface definitions inherent in MOSA, these models have the potential to streamline the verification and validation process. The committee suggests that avionics suppliers be encouraged to construct high-fidelity surrogates for expensive hardware to facilitate system-level testing by including this as a source-selection criterion.

An analogous recertification problem is currently being addressed jointly by the FAA and suppliers of commercial avionics systems, with the emphasis on the reuse and multiple application of software modules. A procedure is being established to give "credit" for full or partial compliance with certification requirements for software components that were originally certified for one application and are targeted for use in a new application. Under the rating/credit arrangement, total-system testing can take "credit" for tests on the reused module, which can concentrate on end-to-end performance requirements and operational safety assessments (FAA, 2000).

The commercial sector has also developed modeling, simulation, and diagnostic tools to ensure the integrity of certification processes and significantly improve the efficiency and shorten the cycle time of avionics testing. The processes established by the FAA and the commercial avionics industry could be used, at least in part, as a model for the Air Force's strategy for requalification testing.

BUSINESS ISSUES

The problem of DMS/OP for avionics maintenance is partly the result of the rapid development of commercial markets for suppliers as new commercial technologies emerged; at the same time, the small defense market dwindled. Inventories of defense-unique parts were naturally depleted, and small orders became unduly expensive as production lines were configured for new markets. Low-volume redesigns of the parts presented huge economic penalties. Thus, both DoD and suppliers were in a losing situation. To cope with this situation today, the Air Force and industry have adopted brute force methods, such as parts substitutions, platform cannibalization, and technology upgrades, as demand requires. Almost everything is done in an unplanned, uncoordinated, reactive way in a budget-constrained environment.

Faced with a somewhat similar situation, the commercial airline industry has resolved the problem through a straightforward analysis of return on investment. When the unreliability of an aircraft or the cost of spares and repairs become unacceptable, and if an alternate solution can be found, the airlines upgrade their avionics components and systems. The future savings are used to justify the cost. In the commercial environment, the supplier is responsible for a significant share of avionics support by providing spare parts throughout most of the service life of the aircraft. In most instances, the avionics supplier has configuration management control, allowing more flexibility in dealing with parts substitution.

The Air Force does not have a comparable business model and the data necessary to make a return-on-investment analysis are dispersed and fragmented. In addition, industry has no incentive to propose alternative solutions to the Air Force's problems. In fact, niche businesses are thriving on the Air Force's dilemma and are content to continue operating with the status quo.

Concerns of Avionics Suppliers

The problems in the business environment related to avionics upgrades are, for the most part, of the traditional, competitive variety. The problems of instabilities, delays, and budgetary constraints are all too familiar in the government/industry business environment. The problem is more complex, however. Avionics suppliers consider upgrades as opportunities to apply their most current technologies and improve their competitive position. Aircraft platform suppliers ("primes") consider upgrades a vehicle for filling in business voids caused by the decrease in new platform opportunities. The tug of war between avionics suppliers and primes intensifies as the supplier wants to maintain its avionics product/system base while the prime attempts to become more vertically integrated to supplement its platform base. DoD/Air Force acquisition practices can affect the "balance" because the buyer determines whether the prime or subsystem supplier is responsible for various upgrade programs. As a rule, normal market forces will resolve these issues, but DoD should consider industrial base issues as part of its acquisition process.

DoD considers upgrades a logical approach to adding functionality and increasing the effectiveness and reliability of avionics systems. The upgrade process is considered an opportunity for introducing new commercial technologies in the form of COTS products and components. Procurements involving pure COTS technology and processes are straightforward. Problems arise when DoD requires changes/modifications/additions to COTS equipment ("COTS plus") to meet unique requirements. The costs (in dollars and human resources) for redesign can be significant; in addition, changes to COTs equipment then raise questions about who owns the new intellectual property.

MOSA will create both opportunities and concerns for avionics suppliers and for DoD. In DoD Directive 5000.2-R, MOSA is defined as follows (DoD, 2001):

> ...a business and engineering strategy to choose specifications and standards adopted by industry standards bodies or de facto standards (set by the market place) for selected systems interfaces (functional and physical), products, practices, and tools.

The committee endorses a strategy that is both engineering based and business based. Various models for avionics procurement are possible with the hierarchical, modular structure of MOSA-based avionics systems. Because of the modular structure, related interfaces will occur at various system levels (e.g., major system, subsystem, functional, line replaceable components of hardware and software). DoD could procure products at any of these levels as avionics systems are upgraded—or even for new systems if architectures are sufficiently predefined (e.g., for the joint tactical radio system). The issue involves choosing the hierarchy level that will support/motivate competition as contrasted with level at which the DoD/service (or prime contractor) might desire to solicit competition. If avionics procurements are mostly at low system-hierarchy levels (e.g., modules) to support incremental upgrades and modernization, the motivation for suppliers of avionics systems and/or avionics "functions" (e.g., flight control, radar, weapons management, etc.) to compete will be relatively low. The high value and competitively leveraging intellectual property (domain knowledge) of these companies is at the system/function level and subsystem level, and they support their R&D investments at these levels. At these higher levels they can also participate in the marketplace with a reasonable return on their investment. New and innovative solutions to avionics needs or requirements are conceived at the system or subsystem level. In the long run, the "system" drives the implementing technologies more than the technologies drive the system. Therefore, the benefit to DoD of competition at the modular level is highly questionable, except perhaps in a few special cases.

The existence of DoD supplier tiers exacerbates the problem. Because fewer military aircraft programs (or major programs in general) are expected in the foreseeable future, airframe suppliers will be motivated to increase their aircraft-maintenance and upgrade business. The tendency of prime contractors to retain more avionics development/manufacturing in house—and the degree of openness of the avionics architecture they propose—will have a major effect on the market (and level of function) available to traditional avionics suppliers and their subtier suppliers. Although government intervention should generally be avoided, these special circumstances dictate that DoD should monitor the developments in the industrial base to maintain at least an awareness of serious potential dislocations.

Intellectual Property

A major business concern that surfaced during this study was the ownership of intellectual property rights in a MOSA environment. The questions of who "owns" the interface standards, who owns the operational or logical information describing and characterizing a replaceable module/object, and who owns the functional avionics architecture embedded in a total system architecture will naturally arise. In most procurements, the real value of the intellectual property rights will usually exceed the "manufacturing cost" associated with a unit. Value-based competition vs. cost-based competition is certainly a related subject, and value-based metrics will be required to support sound business decisions.

Another industry concern is the role of subtier suppliers in a MOSA procurement environment, especially how these suppliers can recover their investments and operate profitably as black-box hardware is replaced by software modules. Business models of these subtier suppliers are normally based on the value of intellectual property and some form of initial payment (the license), coupled with a recurring royalty payment based on use of the property. These models are most pertinent to the intellectual property incorporated in higher volume applications where the costs to develop the intellectual property are high and availability/scarcity and capability normally determine value (Warshofsky, 1994).

As avionics functionality moves away from the traditional black box, subtier suppliers will be faced with redefining their value stream and splitting their traditional hardware into application-specific hardware and applications software, which could be hosted in their own or another supplier's processing cabinet. If the number of sites or platforms is low, subtier suppliers will have to adjust their models to continue to earn a profit. This will require changing the traditional procurement model to include a fair assessment of the subtier supplier's intellectual property rights and a reasonable revenue stream.

The problem is particularly difficult for pricing software modules or "objects." Because much of a supplier's proprietary intellectual property is embedded in software, value-based pricing, rather than cost-based pricing, of software will be necessary to support a sound business model. As indicated above, this will require a change in acquisition practices (and the buying "culture"), particularly in a MOSA environment.

Responsibility for Sustainment

The responsibility for sustaining the combat readiness of military aircraft is currently divided between government and contractor facilities. The details of how this responsibility is divided are important to both parties. The government wishes to maintain in-house expertise in certain core technologies considered critical to the national defense and to sustain minimum workload levels at its facilities. However, because contractors rely on the revenue streams from their maintenance activities to support their business model, the division of maintenance responsibility is an important business issue for them.

The Air Force is required to sustain an inventory of approximately 6,000 aircraft to support 195 active air wings. Maintenance is conducted at three levels: (1) the flight line; (2) intermediate maintenance levels (conducted at the air-wing level); and (3) more extended repairs and upgrades (conducted by contractors or at one of three air logistics centers or depots: Ogden, Warner Robins, or Oklahoma City). Each depot has a large number of hardware and software engineers who design and maintain legacy systems.

As the Air Force relies more heavily on COTS hardware and software and less on Military Specification components in avionics systems, the trend is toward greater reliance on commercial contractors to support these products. To protect the jobs of depot employees, Congress passed USC 2466, Title 10 (the so-called "50/50 rule"), which requires that 50 percent of the depot maintenance workload in certain core technology areas be performed by government employees. This law has had the desired effect of bringing maintenance work that had been contracted out back to the depots but has also raised serious industry concerns.[2]

The depots have the expertise and institutional memory to maintain legacy avionics systems most efficiently, and the committee believes they should continue to do so. However, with increased reliance on COTS and MOSA for major upgrades and new systems, the expertise and intellectual property necessary to maintain these systems will reside more and more in the private sector. Thus, government's role in maintaining these new systems will naturally evolve into overseeing the activities of contractors. Government and industry will have to work together to develop creative solutions to complying with the 50/50 rule in light of changing technological realities. The following solutions could be considered:

- Increase the scope of Air Force logistic centers to include maintenance and modifications of aircraft structures. As the expected lifetime of the aircraft increases, the contractor's experienced workforce in that structure will diminish. For example, if the air logistics centers assume responsibility for performing service life extension programs, they would retain specific aircraft skills and free the contractor to address more of the commercial-based maintenance activities.
- One purpose of the 50/50 rule is to retain jobs at the air logistics centers, thus minimizing negative economic impact to the surrounding community. By giving contractors "credit" for hiring and retaining former depot personnel or hiring subcontractors who retain those skills at on-site or nearby

[2]On August 17, 2000, the Industry Logistics Coalition—a broad-based organization of industry suppliers—wrote to the Secretary of Defense detailing their concerns. The coalition asked Secretary Cohen to hold off on making decisions about bringing Air Force work currently performed in the private sector "in-house" to comply with the "50-50" depot maintenance requirements until a more viable, long-term alternative is developed (letter to Secretary of Defense William Cohen, August 17, 2000).

facilities, the intent of the 50/50 rule would be met. In addition, the economy in local communities would be stimulated.

LOOKING AHEAD

Neither modular nor "open" systems directly address the total DMS problem. However, a MOSA strategy would increase the probability that system components would be available from multiple sources in the future. With proper architecture, planning, and documentation, MOSA could provide an effective strategy for mitigating the problem of obsolescence by ensuring that hardware and software components could be upgraded or replaced over the lifetime of the system. However, the commercial marketplace would determine the need for, and the nature of, publicly open standards beyond those dictated by system-of-system interoperability requirements.

The good news is that the airframe and avionics systems suppliers are already developing and implementing many MOSA-based technical solutions and are working in a loosely coordinated way with DoD to identify and resolve related business issues (see Chapter 3). Avionics systems suppliers are rapidly developing their own modular architectures and modular interfaces. The need for system flexibility and extensibility, coupled with the tremendous competitive leverage of reusing hardware, software, and/or intellectual property, and the need to make more effective use of a limited supply of capable personnel, have forced industry in that direction.

Joint industry/government development of the architecture for the joint tactical radio system, industry interaction with the Open System Joint Task Force, and support from the National Center for Advanced Technologies to the Office for Aging Avionics, Aeronautical Systems Center, and the Open System Joint Task Force for MOSA, are examples of effective industry/government relationships. More coordinated, enterprise-level interactions with industry would be beneficial. By working jointly with industry to resolve business issues, as well as by addressing internal management, budgetary, and technical issues, the Air Force can continue to make progress in mitigating the aging avionics problem.

5

Findings and Recommendations

In this chapter, the committee presents its findings and recommendations for addressing the aging avionics problem. The findings are presented first, beginning with general findings and proceeding to specific findings in the four categories discussed in Chapter 4: management issues, budgetary issues, technical issues, and business issues.

Because the Air Force is the sponsor of this study, the focus of the committee's recommendations is on actions that should be taken by the Air Force. However, during the data-gathering phase of this study, it became apparent that aging avionics is not just an Air Force problem and that all of the services could benefit from a DoD enterprise strategy for dealing with this problem. Therefore, the recommendations are divided into two groups, those that address only the Air Force and those that are applicable to all of the services and should be promoted by the Air Force for implementation at the OSD level. The committee attempted to distill the recommendations down to a few key items that are both important and achievable. There is no one-to-one correspondence between the findings and recommendations.

GENERAL FINDINGS

Finding 1. The problem of aging avionics in military aircraft is large and growing. Unless it is addressed proactively and comprehensively, it will have a significant negative impact on the military readiness of U.S. forces.

The Air Force alone estimates that it needs an additional \$250 million to \$275 million per year to address this problem, and the amount will certainly increase as the rate of obsolescence accelerates. As maintenance requirements and costs increase, fewer aircraft will be in a satisfactory state of readiness and/or mission capable, and less funding will be available for avionics upgrades, resulting in a self-feeding "death spiral."

Finding 2. The amount budgeted for the modernization of Air Force avionics systems is far short of the amount needed to pay for upgrades already approved in critical areas: performance and safety-mandated upgrades; avionics upgrades required for the global air traffic management (GATM) system; and replacements for aging avionics subsystems with the lowest reliability and/or highest repair costs.

The committee estimates that another \$5 billion will be needed beyond FY05 just to complete the upgrades approved in the FY01 President's Budget Request, and these upgrades do not address the entire spectrum of avionics modernization.

Finding 3. A large number of organizations within DoD, the military services, and industry are attempting to address various aspects of the aging avionics problem. However, these efforts are poorly coordinated and often duplicative.

The committee identified more than two dozen organizations in the Office of the Secretary of Defense, the military services, and industry that collectively spend tens of millions of dollars each year on technology development, software tools, manufacturing processes, circuit redesign and reengineering, and policy development to address the aging avionics problem. Although many of these programs are making substantial progress, they are poorly integrated. No enterprise-wide leadership is being provided.

Finding 4. Widespread application of a MOSA to avionics architectures would enable DoD to manage the aging avionics problem more affordably, for both new aircraft and many legacy systems.

Among the many organizations that testified before the committee, there was widespread agreement on this point, although there were diverse interpretations of what MOSA means.

Finding 5. Most of the benefits of MOSA can be realized through a "modular" approach. Although a fully "open" system would have some additional advantages to the government in a few situations (as they do in certain commercial sectors where quantities and related factors can support a viable business case for this approach), most DoD acquisitions cannot justify a totally open approach. The "modular" aspect of MOSA, however, could be applied to virtually all DoD products.

In theory, competition among suppliers in an open-systems approach could reduce government procurement costs. But business models must also be developed to provide incentives for suppliers to participate and to protect the intellectual property rights of avionics suppliers and their subtier sources.

SPECIFIC FINDINGS IN KEY ISSUE AREAS

Government Management Issues

Finding 6. There is no DoD-wide enterprise strategy, and only an embryonic Air Force-wide strategy, for dealing with the aging/obsolescent avionics problem. As a result, no enterprise management or leadership is addressing the problem on a full-time basis.

Partly because of "stove-pipe" management structures, organized around individual weapon systems, management responsibility for dealing with the aging avionics problem is fragmented. The committee found little evidence of cross-program, cross-platform, or cross-service coordination in the Air Force.

Finding 7. The Joint Technical Architecture (JTA) for defining weapon system architectures and standards extends beyond those needed for *inter*platform interoperability. The extension into *intra*platform standards is neither consistent nor integrated with MOSA approaches for addressing aging avionics. In fact, the JTA has shown an alarming reversion to the Military Specification (Mil Spec) era by requiring an onerous number of standards and specifications for *intra*-platform avionics systems.

Finding 8. The technical expertise of DoD's depot support maintenance personnel in state-of-the-art avionics systems appears to be eroding as the workforce ages and retires.

DoD (as well as the defense industry) is having difficulty attracting highly trained young engineers, and many younger, high-potential personnel are leaving government service for industry, where pay scales are higher and opportunities for advancement are more abundant.

Finding 9. As modifications and upgrades of aging avionics systems continue, aircraft, even of the same type, are being equipped with avionics systems with different compositions, capabilities and compatibilities, thus exacerbating the configuration management problem.

Budgetary Issues

Finding 10. Long acquisition and upgrade cycles virtually require that avionics technology-refresh cycles be built into program plans during the engineering and manufacturing development phase prior to initial fielding.

Driven by the commercial market, component product cycles are becoming shorter and shorter, while military acquisition cycles are becoming longer as a result of funding constraints. This mismatch only exacerbates the obsolescence problem and drives up costs.

Finding 11. Because of legal restrictions on the use of appropriated funds in various segregated accounts ("colors of money"), program managers are unable to address aging avionics problems in the most efficient way.

To ensure that appropriated funds are used for their intended purposes, Congress has placed a number of legal restrictions on budget accounts, including accounts available to address the aging avionics problem. Program managers spend a great deal of time trying to manage these color-of-money issues and often lack the budgetary flexibility to address problems logically as they arise.

Finding 12. A comprehensive MOSA solution to the aging avionics problem could save money in the long run but would generally cost more than customized point solutions in the short run. This is particularly true for avionics upgrades in the legacy fleet.

In the current budget-constrained environment, it is difficult to find funding for designs that would lead to reductions in TOCs in the long term.

Technical Issues

Finding 13. Implementation of MOSA would be facilitated by addressing the following needs:

- development of a common understanding of MOSA
- support for development of MOSA building codes, and disciplined design processes and related design tools required for MOSA implementation
- development of a test/requalification strategy coupled with the proper modeling and simulation tools to implement the MOSA strategy economically
- development of design-reuse databases and high-fidelity avionics models by original equipment manufacturers and suppliers

Business Issues

Finding 14. MOSA challenges the traditional military procurement model in several ways:

- With a modular, open-structured avionics system, DoD would, in theory, be able to solicit supplier competition at a variety of systems architecture levels: at the component level, the circuit-board level, the module level, or the subsystem level. The level must be high enough to provide incentives for qualified suppliers to participate, take advantage of local openness, and encourage suppliers to invest in research to improve avionics systems and stimulate innovation.
- The traditional mind-set of acquiring hardware and software will have to be changed to one of acquiring functionality (an approach in keeping with acquisition-reform precepts).
- The protection and value pricing of a supplier's intellectual property will be a key to success and will therefore require workable business models.
- Business incentives must be defined and provided to suppliers that will motivate a MOSA to avionics system design.

Traditionally, the government has acquired a hardware black box (with associated software) to perform a specific function. As signals processors have become more powerful general and digital signal processors, operations previously performed by hardware can now be performed via software algorithms, which are often embedded in the processors of another supplier's black box. The shift toward increasing software content in avionics systems will require a paradigm shift from the procurement of hardware to the procurement of functionality and value.

As the avionics industry moves into the era of software-dominant components and object-oriented design, determining software component pedigrees and ownership will be increasingly difficult. Suppliers must be able to protect their investments in the development of software; at the same time, they must satisfy the government's need for software that is part of a MOSA solution.

Finding 15. As DoD relies more heavily on commercial off-the-shelf hardware and software in avionics systems—and less on Mil Spec components and DoD-unique software languages—the expertise and intellectual property necessary to develop and maintain these systems will increasingly reside in the private sector.

Although the maintenance of legacy avionics systems will continue to be done by personnel at government depots, responsibility for the maintenance of new systems will devolve increasingly onto contractors. Government and industry will have to work together to develop creative solutions that recognize technological realities while complying with government mandates.

RECOMMENDATIONS

The recommendations that follow are divided into in two sections: recommendations that should be implemented internally by the Air Force, and recommendations the Air Force should seek to have implemented externally by OSD to facilitate better management of the aging avionics problem for all of the services.

Recommendations Specific to the Air Force

Recommendation 1. The Air Force, in coordination with the Office of the Secretary of Defense, should develop an "enterprise strategy" for dealing with the aging avionics problem. As a central feature of this strategy, the Air Force should mandate the creation of platform management/upgrade road maps with defined funding requirements for each weapon-system program.

Using these road maps, the Air Force should plan a program and budget for periodic block upgrades of all relevant programs, combining modernization with the resolution of the problem of DMS and the application of evolutionary acquisition principles. Comprehensive road maps to guide individual platform activities would also provide an effective framework for cross-platform and, eventually, cross-service coordination.

Recommendation 2. The Air Force should raise the awareness in Congress about the shortfall in funding for avionics modernization by increasing its congressional budget request to a level consistent with the modernization plans in system road maps.

Program Objective Memorandums should include funding plans for executing program road maps, as described above. Addressing this shortfall will require an increase in appropriations, not just reprogramming of existing funds. Anticipated TOC savings and derived payback periods should be included in the memorandums.

Recommendation 3. The Air Force should require a modular, open-system design strategy for all new programs and upgrades, unless specifically waived. Emphasis should be on achieving the benefits of modularity rather than on complete openness, which often creates business or technical problems. A training program in MOSA concepts should be included for program managers, acquisition personnel, and support personnel. Contractors should be encouraged to use executable specifications as the primary archival documentation of the system; these specifications should be integrated into the avionics design environment.

Recommendation 4. The Air Force should continue to use the Quarterly Acquisition Program Reviews (QAPRs) as a forum for top-level oversight and, most important, for setting priorities to address the aging avionics problem.

Each service has its own management processes for making budgetary decisions and overseeing programs. In the Air Force, QAPRs have previously been used for evaluating the potential of modular open avionics designs to reduce avionics costs.

Recommendation 5. The Air Force software and hardware testing community should develop a testing/requalification strategy tailored to modular avionics systems and should explore methods, including the use of high-fidelity simulation/emulation models and test beds, to minimize the impact on cost and schedule of requalifying avionics components and systems. The Air Force should build on the test strategy and simulation/emulation/diagnostic software model used by the Federal Aviation Administration in the commercial sector, which recognizes the value of reusing hardware/software and provides certification-test credit for reusable modules.

Operational testing and requalification costs and schedules are a major impediment to the smooth incorporation of block changes, especially in legacy equipment. Safety changes should be separated from other types of changes, and detailed emulation and simulation models should be used to minimize actual testing, especially in nonsafety areas. Contractors responsible for executing the changes should be contractually responsible for providing (or making available results from) emulation/simulation models and test histories.

To reduce further the cost/schedule of the testing/requalification process, the Air Force could include funding for the development of related modeling/simulation technology for reducing the need for verification/certification testing in its science and technology budget.

Recommendation 6. The Air Force should examine the feasibility of requiring, as a normal contractual deliverable, contractor-retained high-fidelity avionics simulation models as a means of minimizing validation/certification testing.

Recommendation 7. The Air Force should increase its support for the new Aging Aircraft System Program Office (SPO), in the Aeronautical Systems Center (ASC), by reinforcing its leadership and management responsibility for reducing the total ownership costs of new and legacy avionics systems.

The new Aging Aircraft SPO could become the starting point of an Air Force-wide enterprise strategy for addressing the aging avionics problem. The effectiveness of the Aging Aircraft SPO will depend on how well it is funded and how much support it receives from Air Force leadership.

Recommendation 8. The Air Force should develop and apply innovative contracting approaches that provide incentives for both government and contractors to reduce total ownership costs of avionics systems.

For example, industry could be given complete responsibility for system support (including problems related to diminishing manufacturing sources/out-of-production parts) as part of the initial procurement of a new system or legacy upgrade. Multiyear support contracts could have a similar effect. To stimulate savings in legacy systems, contracting methods that share savings resulting from the introduction of MOSA designs and newer, more reliable technology would be a powerful tool for addressing aging issues. Innovative contract arrangements for government facilities and personnel could be used to ensure compliance with the 50/50 rule.

Recommendations That Apply to All of the Services

Recommendation 9. The Air Force should recommend that the Office of the Secretary of Defense develop an overall "enterprise strategy" for dealing with the aging avionics problem and issue a specific policy directive covering the following four points:

- A modular, open-system design strategy should be required for all new programs and upgrades, unless specifically waived.
- Development and use of program road maps should be mandatory for all Acquisition Category I (ACAT-I) programs (and their use encouraged for lesser programs); road maps should include funding plans and anticipated reductions in total ownership costs.
- Reviews by the Defense Acquisition Board (DAB) of these items should be a required acquisition milestone exit criteria.
- A revolving fund should be established (possibly the Working Capital Fund) to further front-end design/qualification of MOSA-compatible solutions to the problem of diminishing manufactur ing sources.

The MOSA requirement would be analogous to a similar situation in 1994 when the Deputy Secretary of Defense mandated that military specifications be eliminated, unless specifically waived. This resulted in rapid changes in acquisition practices. A similar cultural change could take place with the rapid adaptation of the modular approach in many DoD programs. Administrative procedures that worked well in the past should be used here.

The extent of the road map requirement should be determined after a more detailed review by OSD, but the Air Force has already made an excellent start in implementing road maps. An extension of this process could be applied to all DoD Acquisition Category I (ACAT-I) programs, unless specifically waived, and used for lesser programs as appropriate.

OSD and service oversight will be necessary to ensure adherence to the policy directive, as well as to provide a forum for reviewing waiver requests. Application of MOSA, adherence to platform road maps, and specific reviews of savings should be explicit items in program reviews by the Overarching Integrated Product Team (OIPT) (and service equivalent) program reviews as well as criteria in DAB milestone decisions. Some or all aspects of these mandates might be made Key Performance Parameters (KPPs) or exit criteria for acquisition decision milestones.

In FY00, the Air Force established a parts-obsolescence funding line as part of the Working Capital Fund overhead. However, this line item is relatively small, and additional funding approaches should be considered.

Recommendation 10. The Air Force should recommend that OSD form joint working groups with industry to address policy and business concerns involved in the resolution of aging avionics problems:

- An industry/government steering group should be formed as a focal point for addressing the issues raised by MOSA procurement models and related modifications to the acquisition process, business/competitive models, intellectual property rights, management/pricing, the 50/50 rule, and related issues.
- The role of the Software Engineering Institute (SEI) could be expanded to include the development of MOSA building codes and design tools and processes; SEI could also recommend the process for defining and implementing interface standards at the proper point in the design cycle. The committee believes these changes would be consistent with current plans to reorganize SEI to consolidate software development, system development, and integrated product team (IPT) activities.
- Congress should be encouraged to give DoD managers greater flexibility to shift funds among budget categories to take advantage of opportunities to reduce total ownership costs (TOCs).
- DoD should consider avenues to encourage young people to seek engineering educations focused on embedded software intensive systems and the maintenance of legacy systems.

Resolving aging avionics issues will require close interaction between government and industry, perhaps through established industry associations as a forum for industry to express its views and avoiding corporate conflicts of interest. Numerous examples of past successes are available (notably the National Center for Advanced Technologies Affordability Task Force on Acquisition Reform).

The growing role of software in modern aircraft avionics significantly contributes to rising TOC. Therefore, in addition to funding research to expand the technology envelope, resources must be found to support technology transfer and training for software developers.

Industry and industry associations can play a key role in encouraging Congress to work with OSD to adopt less restrictive constraints on color-of-money and related issues. DoD's adoption of contracting methods that provide greater incentives could encourage industry to become more proactive.

Recommendation 11. The Air Force should recommend that OSD form a joint cross-platform working group (JCWG) at the flag officer level to focus on reducing total ownership cost through the joint development of modular, scalable systems and the use of common solutions across weapon system platforms.

Many systems and subsystems used by different services are faced with the problem of obsolescence. Substantial benefits to all of the services could be derived from the joint development and adoption of common solutions. The end items might not be identical but could be tailored to meet service-unique requirements through the MOSA process, as long as they were derived from a common, modular, scalable avionics family. The JCWG could be organized along the lines of the Joint Aeronautical Commanders Group to leverage common solutions to common problems.

Recommendation 12. The Air Force should recommend that OSD examine and modify traditional defense procurement practices to minimize problems for avionics suppliers.

Significant business problems will arise for avionics suppliers as digital/software-intensive systems replace older systems. Many of these issues will revolve around the ability of suppliers to recover their investments (and make profit) in intellectual property. The profit model currently used would have to be revised to ensure that suppliers would realize a fair return on their investment. OSD should work with industry to explore these issues and modify its procurement practices, as appropriate. For example, intellectual property rights could be retained by suppliers, who would be compensated for their use through value pricing.

Recommendation 13. The Air Force should recommend that the current Open Systems Joint Task Force become the center of expertise and the focal point for addressing issues associated with the application of MOSA. Modularity, rather than total openness, should be emphasized to accommodate current business and technical issues.

Funding for the Open Systems Joint Task Force should be increased to promote the objectives of a MOSA strategy. Staff for the task force may also have to be increased. To clarify that the "modular" aspects of MOSA should be the primary objective, and to minimize confusion associated with the term "open," the name of the task force could be modified to the Modular Open Systems Joint Task Force.

Recommendation 14. The Air Force should recommend that the Office of the Under Secretary of Defense for Acquisition, Technology and Logistics restrict applicability of the Joint Technical Architecture (JTA) and mandated standards to *inter*platform interoperability and allow the *intra*platform standards to be defined by a MOSA approach, along with a greatly reduced number of mandated standards.

Although JTA's promulgation of standards for the interoperability between various platforms is important, in the committee's view, a MOSA strategy for avionics design for a given platform should be based on consensus standards derived from discussions involving both government and industry.

References

Air Force Magazine. 1999. Eight Straight. Journal of the Air Force Association 82(12): 68-70.

Alexander, C. 1977. A Pattern Language: Towns, Buildings, Construction, Oxford, United Kingdom: Oxford University Press.

Borky, J.M., A. Frink, R.N. Lachenmaier, and J.P. Messing. 1998. Architectures for Next Generation Military Avionics Systems. Unpublished white paper handout, IEEE Aerospace Conference, Snowmass, Colorado, March 2000.

CBO (Congressional Budget Office). 2000. Budget Options for National Defense. Washington, D.C.: Congressional Budget Office.

DoD (U.S. Department of Defense). 1999a. DoD Joint Technical Architecture (JTA), Version 3.0. Available on line at: *http://www-jta.itsi.disa.mil/*.

DoD. 1999b. DoD Joint Technical Architecture Development Group Charter. Available on line at: *http://www-jta.itsi.disa.mil/*.

DoD. 2000. The Defense Acquisition System. Directive 5000.1. Washington, D.C.: U.S. Department of Defense, Deputy Secretary of Defense.

DoD. 2001. Mandatory Procedures for Major Defense Acquisition Programs (MDAPs) and Major Automated Information System (MAIS) Acquisition Programs. Directive 5000.2-R (Interim). Washington, D.C.: U.S. Department of Defense, Under Secretary of Defense, Acquisition, Technology, and Logistics.

Donatelli, G. 2000. F-15 Avionics Challenges. Presentation by Lt. Col. Geoffrey Donatelli, Wright Patterson Air Force Base, to the Committee on Aging Avionics in Military Aircraft, Dayton Marriott Hotel, Dayton, Ohio, May 2, 2000.

DSB (Defense Science Board). 1999. Report of the Defense Science Board Task Force on Acquisition Reform, Phase IV. Subpanel on Research and Development. Washington, D.C.: Office of the Under Secretary of Defense for Acquisition and Technology.

Durante, B. 2000. Air Force Funding. Presentation by Blaise Durante, Assistant Secretary of the Air Force, to the Committee on Aging Avionics in Military Aircraft, National Research Council, Washington, D.C., August 17, 2000.

DUSD (AT&L) (Deputy Under Secretary of Defense [Acquisition, Technology, and Logistics]). 2000. Product Support for the 21st Century: A Year Later. Report of the Section 912(c) Study Group for Product Support, September 2000. Washington, D.C.: Deputy Under Secretary of Defense (Acquisition, Technology, and Logistics).

Ebersole, D.C. 2000. Affordable Avionics Solicitation Approach. Presentation by Douglas Ebersole, Aeronautical Systems Center, to the Committee on Aging Avionics in Military Aircraft, Dayton Marriott Hotel, Dayton, Ohio, May 3, 2000.

FAA (Federal Aviation Administration). 2000. Paper on reusable software components. Draft #6 dated June 25, 2000. Certification Authorities Software Team (CAST). #P-40, titled Guidelines for the Reuse of Software Life Cycle Data, dated June 22, 2000.

Hitt, E. 2000. AIAA Brief on Aging Avionics. Presentation by Ellis Hitt, AIAA Digital Avionics Technical Committee, to the Committee on Aging Avionics in Military Aircraft, National Research Council, Washington, D.C., March 27, 2000.

Logan, G.T. 2000. Presentation to NRC on A Vision for Weapon System Electronics Acquisition through the Modular Open Systems Approach to the Committee on Aging Avionics in Military Aircraft, National Research Council, Washington, D.C., March 27, 2000.

NRC (National Research Council). 1997. Ada and Beyond: Software Policies for the Department of Defense. Committee on the Past and Present Contexts for the Use of Ada in the Department of Defense, National Research Council. Washington, D.C.: National Academy Press.

PBR (President's Budget Request). 2000. U.S. Air Force Committee Staff Procurement Backup Book. FY 2001 Amended Budget Request. Aircraft Procurement, Air Force. Volumes I and II. Office of responsibility: Secretary of the Air Force/Financial Management Budget (SAF/FMB). February 2000.

Raggio, R. 2000. Affordable Combat Avionics. Presentation by Lt. Gen. Robert Raggio, Commander, Aeronautical Systems Center, to the Committee on Aging Avionics in Military Aircraft, Dayton Marriott Hotel, Dayton, Ohio, May 2, 2000.

TACTech. 1997. TACTech Data Service Database on obsolescence in military/aerospace devices. Available on line at: *http://www.tactech.com/*.

U.S. Air Force. 1999. Air Mobility Strategic Plan 2000. Scott Air Force Base, Ill: Headquarters Air Mobility Command.

U.S. Air Force. 2000a. Total Ownership Cost (AFTOC) Management Information System. Arlington, Va.: Air Force Cost Analysis Agency.

U.S. Air Force. 2000b. Air Force Fact Sheets. Available on line at: *http://www.af.mil/news/factsheets*.

Warshofsky, F. 1994. The Battle to Own the World's Technology. New York: John Wiley & Sons Inc.

Wasson, J. 2000. Smiths Industries. Presentation by J. Wasson, Component Obsolescence Management for Rotorcraft Avionics Equipment, to the Committee on Aging Avionics in Military Aircraft, National Research Council,. Washington, D.C., June 8, 2000.

Appendix A

Current Activities and Programs

OFFICE OF THE SECRETARY OF DEFENSE

Open Systems Joint Task Force

The Open Systems Joint Task Force (OS-JTF), established by the Office of the Under Secretary of Defense, Acquisition Technology and Logistics (USD[AT&L]), is responsible for providing support for the modular, open-system approach (MOSA), as well as the insertion of commercial open-system technology and products into U.S. Department of Defense (DoD) systems. OS-JTF funds pilot programs and demonstrations with industry to ensure the feasibility of process changes for modular, open systems. OS-JTF is also involved in setting industry standards, international military standards, and determining how they apply to DoD. Working with and providing support for program managers to develop system upgrades or new systems, OS-JTF ensures that lessons learned about modular open systems and pending standards are understood.

Joint Technical Architecture

The Assistant Secretary of Defense (ASD) Command, Control, Communications, and Intelligence (C3I) issued a memorandum on November 14, 1995, to command, service, and agency principals involved in the development of command, control, communications, computers, and intelligence (C4I) systems about addressing the need for joint operations in combat and the reality of a shrinking budget. Recipients of the memorandum were directed to "reach a consensus of a working set of standards" and "establish a single, unifying DoD technical architecture that will become binding on all future DoD C4I acquisitions" so that "new systems can be born joint and interoperable, and existing systems will have a baseline to move towards interoperability" (personal communication with V. Garber, director of interoperability, OUSD [AT&L], November 30, 2000).

The Joint Technical Architecture Group, chaired by ASD(C3I), was formed. Using the U.S. Army Technical Architecture as a starting point, JTA 1.0 Version was released on August 22, 1996, and immediately mandated by the USD (AT&L) and ASD(C3I) for all new and upgraded C4I systems (DoD, 1999).

The development of JTA 2.0 began in March 1997 under the direction of a Technical Architecture Steering Group, cochaired by ASD(C3I) and USD(A&T) OS-JTF. The applicability of Version 2.0 was expanded to include information technologies in all DoD systems (DoD, 1999).

Development of JTA 3.0 began in June 1998. JTA 3.0 includes additional subdomain annexes and incorporates the newly developed DoD technical reference model to ensure that references to standards throughout the document are integrated (DoD, 1999).

Joint Strike Fighter

The Joint Strike Fighter (JSF) Program is a significant new DoD program focused on proactive measures to address the issue of aging avionics in the future. The JSF Program faces three challenges. First, it must develop processes for an affordable, effective, evolvable family of weapon systems with an operational lifetime of more than 30 years, even though most product lifetimes are less than five years. Second, the new weapons systems must be developed according to the system acquisition procedures, which include compliance with acquisition reform processes and the open systems policy; at the same time proprietary information must be protected and the integrity of competition maintained. Third, the new weapons systems must meet the needs of both warfighters (by providing real-time operation, safe flight, and security) and the needs of logisticians (by designs for reliability and maintainability, that facilitate easy upgrades and include plans for dealing with obsolescence). The JSF Program, which is committed to funding a MOSA strategy to meet these challenges, has been collaborating with industry and academia since 1994.

Defense Advanced Research Projects Agency/Information Technology Office Embedded Software

The Defense Advanced Research Projects Agency (DARPA) formed a Tiger Team in the Information Technology Office to determine the potential of improving embedded software in DoD systems. The Tiger Team will address concerns about the growing problem of integrating embedded software in the large, complex systems. Surprisingly, DARPA has determined that the cost of integrating embedded software is 40 to 50 percent of the acquisition cost and growing. Therefore, DARPA is sponsoring studies and pilot programs to establish a new process for reintegrating physical and information sciences. The purpose of these programs is to suggest two new technologies to support this process change.

Defense Microelectronics Activity

The Defense Microelectronics Activity (DMEA) was established by DoD to provide a broad spectrum of microelectronic services. Located in Sacramento, California, DMEA is under the direction and control of the Deputy Under Secretary of Defense for Logistics. Because DMEA is the DoD activity involved with the obsolescence of microelectronics, it has sponsored a number of government and industry initiatives to address the growing problems of sustainment and obsolescence.

DMEA's primary mission is to leverage the capabilities and payoffs of advanced technology to solve operational problems in existing weapon systems, increase operational capabilities, reduce operation and support costs, and reduce the effects of diminishing manufacturing sources (DMS). DMEA assists weapon systems program managers by providing advanced microelectronics technologies, ensuring long-term sustainment of these systems, and providing studies and analyses of current and future sustainment problems.

Diminishing Manufacturing Sources and Material Shortages Teaming Group

The Office of the USD(AT&L) established the Diminishing Manufacturing Sources and Material Shortages (DMSMS) Teaming Group to address the issue of component obsolescence. Members of the DMSMS Teaming Group, who represent DoD programs and industry, are working together to find solutions to common component obsolescence problems. The Teaming Group maintains a database of current information on component obsolescence and, whenever possible, explores resolutions that will work for all programs faced with the obsolescence problem, often reducing the cost. Membership in the DMSMS Teaming Group is open to all procuring activities. Currently, no membership or computer usage fees are required.

Government Industry Data Exchange Program

The Government Industry Data Exchange Program (GIDEP) is a cooperative, government-industry program to reduce costs by making maximum use of existing data. The program provides a medium of exchange for technical information and data essential for research, design, development, production, and sustainment.

GIDEP is managed and funded by the U.S. government and chartered by the Joint Logistics Commanders. Participating organizations include: U.S. Army, U.S. Navy, U.S. Air Force, Defense Logistics Agency, National Aeronautics and Space Administration, U.S. Department of Energy, U.S. Department of Labor, Federal Aviation Administration (FAA), U.S. Postal

Service, National Security Agency, U.S. Department of Commerce, General Services Administration, and the Canadian Department of Defense. Hundreds of commercial companies that produce hardware for the government also participate in the program. GIDEP has been selected by the DoD to be the central repository for DMSMS-related information.

Defense Logistics Agency/Defense Supply Center Columbus

In 1987, the Defense Logistics Agency (DLA) contracted with the Sarnoff Corporation, in collaboration with the Defense Supply Center Columbus, for generalized emulation of microcircuits (GEM) to begin research on replacing older microcircuits, from the earliest form of integrated circuits (ICs) to modern ICs. Using Sarnoff's on-site foundry, computer-aided design software, and comprehensive knowledge of IC design and development, GEM produces circuit replacements that match the form, fit, and function (FFF) requirements of the original part.

The objective of the GEM program is to make available an economical, rapid delivery source of FFF-equivalent microcircuits to support readiness requirements for military equipment. The goal of GEM is to develop a generic emulation system with the capability of supporting on-demand production of microcircuits based on modern technologies.

The GEM alternative provides at least two major benefits. First, it reportedly provides digital microcircuits at approximately one-tenth the cost and in one-quarter the time required for developing and testing a redesign alternative. Second, the GEM system design includes a reuse strategy to ensure that design and fabrication building blocks are catalogued, promote cost containment (i.e., least cost for multiple users), and ensure the long-term availability of families of microcircuit devices (personal communication with Justine Corboy, Sarnoff Corporation, July 25, 2000).

Shared Data Warehouse

The DMS Shared Data Warehouse is being developed by the Division of Industrial Support Program (DLA) to enable DoD to manage parts obsolescence. The objective of the DMS Shared Data Warehouse is to minimize the impact of DMS on DoD weapon systems. The Shared Data Warehouse promotes a systematic, single methodology for processing notices of discontinuance and provides a central repository for DMS management. Using business-type process evaluations, in addition to existing screening processes, the DMS Shared Data Warehouse provides rapid, economical identification, dissemination, and processing of affected part numbers and national stock numbers.

AIR FORCE

Aeronautics Systems Center's Affordable Combat Avionics

The Air Force received an action item from the Quarterly Acquisition Program Review in October 1998 to present a plan for studying the design of avionics systems to preclude their obsolescence. The Affordable Combat Avionics Office has been the single most active Air Force organization addressing the problem of aging avionics (Raggio, 2000). First, they are attempting to institutionalize open-system avionics architectures by providing policy guidance and direction to program managers. Second, they are sponsoring studies with industry and working with original equipment manufacturers to identify challenges and exchange ideas. Third, the chief architect is establishing a process for integrated-change road maps to identify opportunities for pilot programs.

The Affordable Avionics Initiative, under the authority of Aeronautic System Center and located at Wright Patterson Air Force Base, has done an excellent job of promoting activities to reduce the total ownership cost (TOC) of new and legacy military avionics systems. It has initiated related discussions and studies with industry, is working closely with the OSD OS-JTF, and is becoming a recognized focal point for aging avionics issues in the Air Force. However, its effectiveness in dealing with current DMS problems and weapons-system modernization has been hampered by the lack of an Air Force (and DoD) enterprise strategy, as well as by the current independent program management structure.

To correct these problems, the Air Force has placed the Affordable Avionics Initiative under the authority of the recently created Aging Aircraft System Program Office (SPO), which will be headed by a general officer. The charter for this office includes not only aging avionics, but also all aspects of force sustainment with the aging inventory (i.e., aging airframes, engines, support equipment, and training systems).

Diminishing Manufacturing Sources and Material Shortages Program

The Air Force Materiel Command (AFMC) asked the Air Force Research Laboratory (AFRL) to provide DMSMS support for weapons systems in its command (AFMC, 2000). AFRL established a DMSMS hub to reduce the impact of obsolescence by distributing discontinuance notices, developing tools and databases for identifying and resolving problems, and ensuring the continued availability of parts to support requirements for current and planned weapon systems. The DMSMS Hub has also been chartered to develop more centralized, coordinated DMSMS management, including support for tracking databases, developing new tool capabilities, and training.

The DMSMS Hub collaborates on numerous OSD and Air Force-level DMSMS-related programs and issues, including initiatives on aging avionics, aging aircraft, open-systems architecture, and operational, safety, suitability and effectiveness. The AFMC DMSMS Hub's Internet web site (*http://www.ml.afrl.af.mil/ib/dpdsp/dmsms.htm*) provides important information, policies and regulations, a calendar of upcoming and recent events, tools and publications, information about the AFMC DMSMS Teaming Group, and links to other DMSMS-related web sites and Air Force DMSMS-related initiatives and focal points.

Electronic Parts Obsolescence Initiative

The AFRL, Materials and Manufacturing Directorate, Manufacturing Technology Division, Wright-Patterson Air Force Base, has implemented a five-year, $21 million initiative to deal with parts obsolescence and the Application of Commercially Manufactured Electronics (BAA-98-14-MLKT) (Bumbalough, 2000). In addition, contractors are providing more than $11 million for this initiative. The initiative consists of eight programs in the following areas: (1) commercially available obsolescence-management decision and reverse-engineering tools; (2) application of commercially manufactured electronics to address key technology-driven issues at the chip, board, and box level; and (3) pilot programs to improve business policies and obsolescence-management processes by using tools and technologies from other areas and to demonstrate and document the cost effectiveness of implementing them into weapon systems.

NAVY

Naval Aviation Systems Team

The Naval Aviation Systems Team is addressing the issue of aging avionics through proactive studies and pilot programs to prevent the problem from affecting the readiness and availability of Navy aircraft. The Navy has established an enterprise team to study aging avionics in legacy systems. The team looks for ways to use open-systems designs that do not require major system upgrades and to continue the Navy's tradition of using common avionics systems to reduce TOC. An organizational problem for the Naval Aviation Systems Team is that program managers are generally assigned life-cycle responsibility for a program but have no dedicated budgets for monitoring or managing problems associated with aging avionics (J. Johnson, 2000).

Helicopter Modernization Program

The Helicopter Modernization Program was established several years ago. A system being developed by Lockheed Martin will replace four legacy platforms in the current inventory with one basic platform (Sikorsky's H-60 helicopters) adapted to perform unique missions. The new platform will have a reusable, open-system architecture with common hardware in the cockpit and common software modules, configured to meet mission needs. The architecture will be compliant with the JTA. The Navy expects to realize a 57 percent reduction in flyaway costs and a 60 percent reduction in TOC (J. Johnson, 2000).

Diminishing Manufacturing Sources Technology Center

The Navy established a DMS Technology Center (DTC) through the Naval Surface Warfare Center, Crane Division, to produce and distribute product discontinuance notifications to those who sign up for this free service. The DTC has the capability of analyzing weapons system bills of materials and producing health assessment reports and supplemental support analyses using the technology-obsolescence, risk-assessment model (TORA), as well as analyses of solution alternatives and recommendations, with corresponding cost analyses.

ARMY

Aviation Electronic Combat Office

The Aviation Electronic Combat Office is responsible for centralized avionics management for Army aircraft systems. The office is implementing the Army's strategy for aging avionics by fostering an open-systems architecture environment and by improving obsolescence management. Using MOSA, the Army is replacing legacy systems through the modernization of major platforms. In addition, horizontal technology integration initiatives, such as the joint tactical radio system (JTRS), will provide common solutions to improve platform performance, reduce size and weight, mitigate against obsolescence, and reduce TOC.

Modernization Through Spares/Continuous Technology-Refresh Program

Obsolescence management is being addressed through preplanned product improvement programs that emphasize reliability and technology-transition plans to reduce TOC. Initially called the Modernization Through Spares Program, the program is now the Continuous Technology-Refresh Program. The rationale for the acquisition of spares is to replace obsolescent parts, maintain performance levels, and avoid failures caused by a lack of parts. Savings will be reinvested in future programs (Johnston, 2000).

Aviation Applied Technology Directorate

The Aviation Applied Technology Directorate (AATD) establishes science and technology strategy for the aviation fleet. Obsolescence and DMS have impacted not only aging aircraft, but also the Apache Longbow and Comanche platforms, which have not yet been fully fielded. AATD is addressing these problems in several ways, notably through the Rotary Open-System Architecture (ROSA) Program and ROSA-D, a demonstration program for the ROSA technology (D. Johnson, 2000).

Rotary Open-System Architecture Program

The objective of the ROSA Program is to select, develop, and evaluate key components of an open-system avionics architecture for dual application to military and civilian rotorcraft. The goal is to create a rotorcraft technical architecture for Army aircraft that would provide descriptions and design standards for high-speed networks, integrated processors, and other commercial off-the-shelf (COTS)-based components. The ROSA-D Program, which is currently unfunded, would demonstrate the applicability and feasibility of concepts developed under the ROSA Program (D. Johnson, 2000).

AEROSPACE INDUSTRY

Lockheed Martin

Proven Path

The objective of Proven Path is to bring together the best technologies and business practices throughout the company to address DMS, technology-refresh strategies, and COTS-based technology applications. The Office of the Corporate Vice President of Technology is responsible for Proven Path. All major business sectors are involved in the Proven Path initiative (Frew, 2000).

Systems, Technologies, Architecture, and Acquisition Reform

Lockheed Martin has been working with the Air Force Aging Avionics Office at Wright Patterson Air Force Base to study the problem in detail. The Systems, Technologies, Architecture, and Acquisition Reform (STAAR) Study, which involves business groups from 10 Lockheed Martin company locations, focuses on several Air Force platforms, including the F-16, F-22, F-117, C-5, and A-10. The final report will provide integrated technical and programmatic solutions to reducing the TOC for each platform. In addition, the report will recommend a business concept for leveraging cross-platform investments to enable integrated technical and programmatic solutions (Sarama, 2000).

The Boeing Company

Bold Stroke

Bold Stroke is a company-wide, company-funded avionics-affordability initiative. The program was started in 1996 at the Phantom Works but cuts across all three of Boeing's major business areas: space, military aircraft, and commercial aircraft. The objective of

Bold Stroke is to use commercially available computer technologies to provide military avionics systems for half the development cost, half the flyaway cost, and less than half the support costs of current systems. The architecture uses a layered software infrastructure to provide hardware/software isolation. Bold Stroke architecture is based on open standards that are available to all suppliers and have no proprietary "hooks" (Varga, 2000).

Open Avionics Systems Integration Study

Boeing has been working with the Air Force Aging Avionics Office on a no-cost-to-the-government study, the Open Avionics Systems Integration Study (OASIS). The study was undertaken to examine five aircraft weapon systems in detail to develop a multiplatform, open-systems solution to rapid, affordable modernization and to lower TOC. This study was conducted by Boeing company organizations representing the B-1B, B-2 (with the participation of Northrop Grumman), B-52, C-17, and F-15. The OASIS strategy began with five programs with individual road maps, evolved to the development of multiplatform integrated road maps focused on a business case-analysis process, and concluded with the development of an affordable migration strategy with open-systems architectures that would lead to the development of common, open-systems, multiple-system migration plans (Seal, 2000).

Commercial Product Offerings for Obsolescence Management

Avionics Component Obsolescence Management

The Avionics Component Obsolescence Management (AVCOM) tool, initially developed under an Air Force contract in support of the F-15 program at Warner Robins Air Force Base, is provided by Manufacturing Technology, Incorporated (MTI). Air Force participation in the development of AVCOM and its refinements have helped structure one of the best DMSMS programs in the Air Force (personal communication with Mike Amspacker, Manufacturing Technology, Inc., August 16, 2000).

AVCOM uses component information from the indentured parts breakdown of the technical order and the source control drawing describing the approved components for use in the weapon system. Once the full system structure has been loaded, any system, box, or board in this hierarchical structure can be selected and analyzed. Analyses include a listing of all next-lower assemblies, a listing of any parts that have known discontinuance announcements, a projection of part availability, and customized queries to provide custom reports. The manufacturing and availability status of all components, as well as all military-equivalent parts associated with systems in the AVCOM application, are monitored by MTI.

Transition Analysis of Component Technology, Incorporated

Transition Analysis of Component Technology, Incorporated (TACTech), is a commercial, interactive, data service that provides internally developed software and parts libraries to address obsolescence problems for semiconductors. Founded in 1987, TACTech provides component life-cycle data to DoD and more than 100 companies worldwide. TACTech's database identifies potential sourcing problems and provides projections for addressing the problem of parts obsolescence proactively. TACTech can analyze systems bills of materials and provide FFF replacement recommendations via a real-time electronic library. With TACTech's indenturing capability, analysis can be done at any level in the weapon system. Base-part numbers or generic-part numbers and parametric-part searches can be done to provide alternatives. TACTech can link users in a teaming environment to enable coordinated decision making and cost-sharing opportunities (TACTech, 2000).

The B-2 program, a team project involving ARINC, DMEA, TACTech, and Northrop Grumman, has been praised for its aggressive, proactive DMSMS program and has been singled out as a model Air Force program (B-2, 2000).

SMART Parts

The goal of SMART Parts, a program under Litton-TASC Corporation, is to reduce TOC through an innovative hardware architecture and design process that provides a brand new solution to the long-term sustainment of aging digital systems (Abrahamson, 2000). One of the most creative features of this innovative architecture is the concept of dynamic adaptability. SMART Part designs are FFF replacements for existing circuit assemblies (CCAs). A SMART Part design can be used in a system without any system modifications and can be installed into multiple CCA locations. The function of the card adapts to meet the system requirements of the CCA it is replacing by simply setting a configuration switch to the desired function. In its purest form, the SMART Part approach could replace an entire line replacement unit (LRU) of unique CCAs with multiple instances of the same SMART Part design. The SMART Part approach works well for purely digital designs.

The SMART Part reimplementation methodology is based on the premise that the boards being replaced meet operational performance requirements but are not supportable for other reasons. By focusing on reimplementing existing designs, identical functionality can be achieved much faster than through a total redesign.

If the existing design does have functional problems and the existing function must be modified, the SMART Part design methodology also supports total redesign. SMART Part designs use "generic" components; the unprogrammed SMART Part circuit card is, in essence, a "blank slate." Firmware defines the functionality. Because there are no application-specific components in the design, SMART Part component replacements are available indefinitely. Updating the SMART Part design to accommodate new components is a comparatively quick and inexpensive task (Abrahamson, 2000).

National Rotorcraft Technology Center

The National Rotorcraft Technology Center (NRTC) is an interagency team from the National Aeronautics and Space administration (NASA), the Army, the Navy and the FAA. The NRTC, located at NASA Ames Research Center, Mountain View, California, cooperatively develops and implements dual-use rotary-wing technology that addresses both civil and military needs. The goal is to ensure the continued superiority of DoD rotorcraft systems while providing an additional dual-use benefit, thus improving the U.S. rotorcraft industry's competitiveness in the civil sector. The NRTC has been the catalyst for a paradigm shift to a new way of doing business between government and industry emphasizing cooperation, streamlined processes and minimum infrastructure (Morris, 2000).

The NRTC's primary program is an innovative approach to including U.S. industry and academia as partners through a focal point, the Rotorcraft Industry Technology Association (RITA), a nonprofit corporation formed for this purpose. RITA is jointly managed and executed. Industry provides at least 50 percent of the funding for all projects; government funding is provided to execute the program through a funded cooperative agreement established under the NASA Space Act. Technology needs are identified by the customer and are strongly focused on dual use. Projects are defined by industry (RITA) in consultation with the government. A federated approach to sharing of facilities and expertise is emphasized, and the NRTC Government Office facilitates access to government laboratories and capabilities. Research data and rights are shared among RITA members.

Software Engineering Institute

The Software Engineering Institute (SEI) is a federally funded research and development center sponsored by DoD through the Office of the USD(AT&L). DOD established SEI in 1984 via a competitive award to Carnegie Mellon University to advance the practice of software engineering to ensure that quality software could be produced on schedule and within budget (DoD, 2000).

The SEI mission is to provide leadership in advancing the state of the practice of software engineering to improve the quality of systems that depend on software. SEI promotes the evolution of software engineering from an ad hoc, labor-intensive activity that is well managed and supported by technology. SEI has been instrumental in developing the Capability Maturity Model, which measures an organization's process capabilities to produce quality software. SEI focuses on two principal areas:

- software engineering management practices, the ability of organizations to predict and control quality, schedule, cost, cycle time, and productivity when acquiring, building, or enhancing software systems
- software engineering technical practices, the ability of software engineers to analyze, predict, and control selected properties of software systems, and make key choices and trade-offs when acquiring, building, or enhancing software systems

SEI has been actively involved in work on open systems since 1993 by developing courses, related products, and other sources of open-systems information, and working on formal standards. SEI work on software architectures has focused on the following areas: architectural evaluation techniques based on attribute-specific models; architecture reconstruction of a specific system implementation as a means of checking a system to ensure that it complies with the architecture specified for it; and to providing information on architectures and architectural concepts to stakeholders.

National Center for Advanced Technologies

In 1993, the National Center for Advanced Technologies (NCAT) formed the Industry Affordability Task Force as part of the Affordability Thrust of the Director of Defense Research and Engineering and DARPA. The task force has evolved into the Multi-association Industry Affordability Task Force, which addresses common industry and DoD science and technology, manufacturing, product and process development, commercial integration, sustainment, and defense acquisition policy and program issues.

Many successful projects featuring the coordinated efforts of various industry associations, government agencies, and key defense industry producers have been co-sponsored and facilitated by NCAT. These projects, performed by senior and executive-level industry experts, have been conducted through workshops, symposia, conferences, and team activities involving one-on-one, face-to-face interviews and group workshops. NCAT's industry experts have dealt with a variety of advanced technology, manufacturing infrastructure, and managerial technology issues in aerospace, aeronautics, avionics, electronics, propulsion, and materials manufacturing, including management technologies (NCAT, 2000).

The Air Force has solicited NCAT to support and facilitate government-industry interactions in support of the Affordable Avionics Initiative for the AFMC. Because of NCAT's affiliation with the professional and industry associations involved in the Multi-association Industry Affordability Task Force, NCAT can provide industry insight into complex issues. NCAT's structure and charter have positioned it to provide rapid responses to industry/policy issues, such as those under scrutiny by the AFMC.

Computer-Aided Life-Cycle Engineering Electronic Products and Systems Center, University of Maryland

Computer-Aided Life-Cycle Engineering Center (CALCE) is sponsored by more than 50 commercial and government organizations from all facets of the electronics systems industry. Over the last 15 years, CALCE has invested more than $50 million in developing methodologies, models, and design tools to address the design and manufacturing of electronic systems (CALCE, 2000). CALCE is recognized as a founder and driving force behind the development and implementation of physics-of-failure approaches to reliability and life-cycle prediction, as well as a world leader in accelerated testing and electronic parts selection and management. CALCE has chaired the development of several reliability and part-selection standards and is at the forefront of the development of international standards for critical electronic systems.

Located at the University of Maryland, CALCE brings together leading avionics, automotive, computer, semiconductor, and electronics manufacturers by providing information and services that match industry needs and an organizational structure in which different sectors of the electronics-industry supply chain can share information and influence practices and policies.

CALCE has been asked by avionics, airframers and end users to outline an avionics road map and transition it to industry (CALCE, 2000). The avionics road map project will address the following issues:

- chart external pressures compelling avionics manufacturers to design with commercial-grade components and subsystems
- identify necessary changes in design, support, and certification
- identify necessary changes in governing regulations and standards
- suggest ways avionics manufacturers can maximize their ability to respond quickly to obsolescence

REFERENCES

Abrahamson, C. 2000. SMART Parts. Presentation by Carl Abrahamson, Litton/TASC, to the Committee on Aging Avionics in Military Aircraft, National Research Council, Washington, D.C., June 8, 2000.

AFMC (Air Force Materiel Command). 2000. Available on line at: *http://www.ml.afrl.af.mil/ib/dpdsp/dmsms.htm.*

B-2 Stealth Bomber. 2000. Proactive DMSMS Management in the B-2 System Progam Office. Material provided to the Committee on Aging Avionics in Military Aircraft, Dayton Marriott Hotel, Dayton, Ohio, May 2, 2000.

Bumbalough, A. 2000. Electronic Parts Obsolescence Initiative. Presentation by Anthony Bumbalough, Air Force Research Laboratory, to the Committee on Aging Avionics in Military Aircraft, Dayton Marriott Hotel, Dayton, Ohio, May 3, 2000.

CALCE (Computer-Aided Life-Cycle Engineering Center). 2000. Electronic Products and Systems Center. Available on line at: *http://www.CALCE.UMD.EDU/general/center/Center/center.htm.*

DoD (U.S. Department of Defense). 1999. DoD Joint Technical Architecture (JTA) Version 3.0. Available on line at: *http://www-jta.itsi.disa.mil/.*

DoD. 2000. Software Engineering Institute, Carnegie Mellon University. Available on line at: *http://www.sei.cmu.edu/.*

Frew, R. 2000. Systems, Proven Path. Presentation by Russell Frew, Lockheed Martin, to the Committee on Aging Avionics in Military Aircraft, National Research Council, Washington, D.C., June 8, 2000.

Johnson, D. 2000. Aging Avionics Science &Technology Strategy and Program Status. Presentation by Dale Johnson, Army Aviation Applied Technology Directorate, to the Committee on Aging Avionics in Military Aircraft, National Research Council, Washington, D.C., June 8, 2000.

Johnson, J. 2000. Naval Aviation Systems Team. Presentation by James Johnson, Naval Air Systems Command, to the Committee on Aging Avionics in Military Aircraft, National Research Council, Washington, D.C., March 27, 2000.

Johnston, L. 2000. Aging Avionics. Presentation by Larry Johnston, Aviation Electronic Combat, to the Committee on Aging Avionics in Military Aircraft, National Research Council, Washington, D.C., June 8, 2000.

Morris, C. 2000. Rotocraft Center of Excellence and Rotorcraft Industry Technology Association. Presentation by Charles Morris, National Rotorcraft Technology Center, to the Committee on Aging Avionics in Military Aircraft, National Research Council, Washington, D.C., August 17, 2000.

NCAT (National Center for Advanced Technologies). 2000. MOSA. Meeting by the MOSA Industry Steering Group, National Center for Advanced Technologies, Washington, D.C., November 1, 2000.

Raggio, R. 2000. Affordable Combat Avionics. Presentation by Lt. Gen. Robert Raggio, Commander, Aeronautical Systems Center, to the Committee on Aging Avionics in Military Aircraft, Dayton Marriott Hotel, Dayton, Ohio, May 2, 2000.

Sarama,T.J. 2000. Systems, Technologies, Architectures and Acquisition Reform (STAAR) Report. Presentation by Thomas J. Sarama, Lockheed Martin, to the Committee on Aging Avionics in Military Aircraft, Dayton Marriott Hotel, Dayton, Ohio, May 3, 2000.

Seal, D.W. 2000. Open Avionics Systems Integration Study (OASIS). Presentation by Daniel W. Seal, the Boeing Company, to the Committee on Aging Avionics in Military Aircraft, Dayton Marriott Hotel, Dayton, Ohio, May 3, 2000.

TACTech. 2000. TACTech Data Service Database on obsolescence in military/aerospace devices. Available on line at: *http://www.tactech.com/.*

Varga, R. 2000. Bold Stroke. Presentation by Reg Varga, The Boeing Company, to the Committee on Aging Avionics in Military Aircraft, National Research Council, Washington, D.C., June 9, 2000.

Appendix B

Biographical Sketches of Committee Members

Robert L. Cattoi (chair) is a retired senior vice president of research, engineering and manufacturing processes at Rockwell International Corporation. In November 1991, Mr. Cattoi was appointed chairman of the U.S. delegation to the International Steering Committee on Intelligent Manufacturing Systems (IMS) the goal of which to systematize, standardize, and develop manufacturing science and technology to provide a basis for agile and intelligent manufacturing systems in the twenty-first century. Mr. Cattoi earned a B.S. in electrical engineering from the University of Wisconsin. He also is a registered professional engineer in the state of Texas and a member of the National Society of Professional Engineers.

Noel Longuemare (vice chair) is a private consultant. In his tenure at the U.S. Department of Defense (DoD) he was Principal Deputy Under Secretary of Defense for Acquisition and Technology for four years and Acting Under Secretary of Defense for Acquisition and Technology for two six-month periods. He was responsible for all matters relating to DoD acquisition. Mr. Longuemare was a corporate vice president and general manager of the Systems Development and Technology Divisions, Westinghouse Electronic Systems Group where he played a leading role in the development of modern radar and avionics systems for airborne and land-mobile applications. He is a fellow of the Institute of Electrical and Electronics Engineers (IEEE) and the American Association for the Advancement of Science (AAAS) and an associate fellow of the American Institute of Aeronautics and Astronautics (AIAA). A registered engineer in Maryland, Mr. Longuemare graduated from the University of Texas-El Paso with a B.S.E.E., Johns Hopkins University with an M.S.E., and the Stanford University Executive Program.

Henry P. Arnold is the president of BF Goodrich Aerospace, Fuel and Utility Systems, headquartered in Vergennes, Vermont. He received an undergraduate degree in electrical engineering from Seattle University in 1972, an M.B.A. from Seattle University in 1985, and an Executive M.B.A. from Massachusetts Institute of Technology in 1989. In his 25 years at Boeing Commercial Airplane Group, he rose to the position of executive vice president of airplane development and definition. He also served as chief project engineer for the 737 Next Generation Program and director of commercial avionics systems. His areas of expertise include product development and program management.

William C. Bowes is vice president of programs management for Litton Industries' integrated systems division. He received a B.S. in chemical engineering from the University of Idaho and an M.S. from the Naval Postgraduate School, Monterey. He also graduated from the U.S. Navy Test Pilot School. While in the

Navy, he was promoted to the rank of vice admiral and served as the commander of the Naval Air Systems Command and Principal Deputy Assistant Secretary of the Navy (RD&A). His areas of expertise include naval-aviation systems engineering, life-cycle management, and program management. He is a member of AIAA, Society of Experimental Test Pilots, Association of Unmanned Aerial Vehicle International and Association of Naval Aviation. His expertise is in mili tary support (operations and maintenance).

Stephen N. Buss is the manager of sustainment initiatives for the Electronics Sensors and Systems Sector of Northrop Grumman Corporation. He is a recognized leader in parts obsolescence and diminishing manufacturing sources strategies, corporate program manager for the Defense Microelectronics Activity Advanced Technology Support Program, and program manager for an Air Force program investigating the use of commercially manufactured electronics in military weapons systems. He is also the program manager for the Common Circuit-Card Assembly Program, a proof of concept program targeted at developing a generic circuit card that can be programmed to perform functions of many obsolete boards. Since 1991, he has proactively created and implemented best practices for component engineering and component procurement and assisted in the implementation of a component-supplier management system. He was an original member of the DoD Producability and Supportability Working Group and the spin-off teaming group. He was committee chair for the Air Force Research Laboratory Diminishing Manufacturing Sources and Material Shortages Hub Users Group Subcommittee, and a member of a consortium for the Aging Avionics/ Electronics Initiative. He has a B.S. in business administration from Towson State University. Mr. Buss has made numerous presentations on diminishing manufacturing sources and mitigation of parts obsolescence.

John D. Cosgrove retired from Rockwell Collins in 1999, where he was president of the company, as well as a corporate officer and senior vice president of Rockwell International. Previous to that, he had been president of Collins Avionics and Communications Division. Mr. Cosgrove was a member of the National Security Telecommunications Advisory Committee and a member of the Armed Forces Communications and Electronics Association. He has a B.S. in electrical engineering from Iowa State University and is a member of the Iowa State University Foundation's Board of Governors. His expertise is in electronics and electrical/industrial engineering.

Frederick H. Dill, a member of the National Academy of Engineering (NAE), is a member of the senior technical staff for the IBM T.J. Watson Research Center. He received his Ph.D. in electrical engineering from Carnegie Institute of Technology in 1958 and was elected to the NAE in 1990 for his pioneering accomplishments in microelectronics technology. As a member of the technical staff for IBM Corporation Research Division from 1958 to 1963 working on exploratory devices, he built the first tunnel diodes and injection lasers in IBM and is the part owner of IBM patent for the injection laser. He was IBM Research Division Manager of high-speed integrated circuit research from 1963 to 1968. This program pitted germanium with its higher mobilities against silicon technology. Following a year as visiting lecturer at University of California, Berkeley, senior and graduate courses, he was manager of IBM Research Division groups working on optical lithography and semiconductor process measurement. Mr. Dill served as IBM Corporation Research Division Senior Member of the Technical Staff working on application of computers in semiconductor manufacturing, factory floor control systems, and multitool process control loops. Mr. Dill is an acknowledged leader in the field of microelectronics technology and engineering science.

Llewellyn S. Dougherty is the director for technology for Raytheon Systems Company. He has served in other areas of the company, including sensors and communications, radar systems and reconnaissance systems. Previous to Raytheon, he was technical assistant to the Director of the Defense Advanced Research Programs Agency (DARPA). His areas of expertise include avionics, digital computers, software, systems engineering and systems safety. A member of IEEE, he earned a B.S. in astronautics and engineering sciences from the U.S. Air Force Academy, an M.S. in aeronautics and astronautics from Massachusetts Institute of Technology, and a Ph.D. in digital systems engineering from the Air Force Institute of Technology.

Valerie J. Gawron is a Level 5 (world-class) engineer at Veridian Engineering Flight Research Group. Her experience in engineering psychology and human factors covers the areas of design, research, simulation, and training. She has produced numerous simulation programs and training manuals to improve aviation. She is a member of the Aerial Space Human Factors Association, Aerial Space Medical Association, and Association of Aviation Psychologists, an associate fellow of the AIAA, and a fellow of the Human Factors and Ergonomics Society. She has a B.A. in psychology, an M.S. in industrial engineering, an M.B.A. from the State University of New York at Buffalo, an M.A. in psychology from the State University College at Geneseo, and a Ph.D. in psychology from the University of Illinois. She has taught at New Mexico State University, University of Illinois, and State University College at Geneseo. Her expertise is in human factors and avionics testing.

David R. Heebner, a member of the NAE, is the proprietor of Heebner Associates. In 1992, he retired from Science Applications International Corporation (SAIC) and became a private consultant. While at SAIC, Mr. Heebner was executive vice-president and vice chairman of the Board of Directors and supervised a multigroup organization that included both the Military Sciences Group and the Information Systems Group. Before joining SAIC, he was deputy director for tactical warfare programs under the Director of Defense Research and Engineering at DoD. Prior to that, Mr. Heebner spent more than 16 years with Hughes Aircraft Company. He currently serves on several boards of directors, has chaired the Naval Studies Board of the National Research Council, and is an active member of the Defense Science Board. He received a B.S. in electrical engineering from Newark College of Engineering and an M.S. in electrical engineering from the University of Southern California. His expertise is in military procurements and controls.

Ellis F. Hitt is a senior manager for Battelle Corporation and the chairman of the AIAA Digital Avionics Technical Committee. He has a B.S. in electrical engineering from the University of Kansas, an M.S. in electrical engineering from the Air Force Institute of Technology, and pursued post graduate studies from Ohio State University and the University of New Mexico. Mr. Hitt is a nationally recognized authority on avionics and flight control systems. He has extensive experience in conceptual, preliminary, and final design of avionics, including navigation, guidance, control, communications, controls and displays, sensors, stores management, weapons delivery, and electrical power subsystems; integration, testing, and analysis of avionics; development of mathematical models and computer programs for performing error analysis, systems simulation and evaluation, and life-cycle cost analyses; mission software design, development, validation, and verification. Mr. Hitt's current responsibilities include senior marketing manager for the Air Force market sector and technical leader on total ownership cost.

Andrew J. Kornecki graduated from the University of Mining and Metallurgy in Krakow, Poland, with an M.S.E.E. and Ph.D. in system engineering. He is currently a professor in the Department of Computing and Mathematics of Embry Riddle Aeronautical University (ERAU) in Daytona Beach, Florida. His areas of expertise include software construction for real-time, embedded, safety critical systems, computer simulation and aviation software, and control and computer engineering.

Rocky J. Porzio has worked for Federal Systems for more than 30 years, where he has been involved in virtually all major elements of systems integration. He is currently the director of avionics systems engineering, responsible for the technical performance of fixed and rotary wing aircraft and technology programs. He has served the company in a variety of positions, including chief engineer and acting director of new business pursuits. He received a B.S.E.E. from the University of Detroit.

George W. Sutton is currently a principal engineer for ANSER Corporation, supporting the Ballistic Missile Defense Organization/Air Force Space-Based Laser Project. He received a B.S.M.E. from Cornell University, an M.S.M.E. from California Institute of Technology, and a Ph.D. in mechanical engineering and physics from California Institute of Technology. He was elected to the NAE in 1994 for contributions to ballistic missile reentry, lasers, medical devices, imaging systems, and aerooptics. He is a member of ASME and a fellow of AIAA. His awards and honors include: U.S. Air Force Outstanding Service Award, 1965; Arthur D. Flemming Award, 1965; Fellow, AAAS; Fellow, AIAA; Thermophysics Medal, AIAA, 1980; and AIAA Outstanding Achievement Award, 1990. He is an exceptionally knowledgeable generalist in aerospace engineering.

William G.T. Tuttle, Jr., General, U.S. Army (retired), has been president and chief executive officer of the nonprofit Logistics Management Institute since January 1993. As the Army's senior logistician, General Tuttle led 100,000 soldiers and civilians of the U.S. Army Materiel Command from 1989 until his retirement early in 1992, a period encompassing Operation Just Cause in Panama, Operation Desert Shield, and Operation Desert Storm. General Tuttle also commanded the U.S. Army Logistics Center (now the Combined Arms Support Command), the U.S. Army Operational Test and Evaluation Agency, the Eastern Area of the Military Traffic Management Command, and both the Support Command and Supply and Transport Battalion of the 3d Armored Division in Germany. He served in the Pentagon as the Army's Director of Force Management and at Supreme Headquarters Allied Powers Europe as Chief of Policy and Programs Branch and representative to NATO's Defense Review Committee. A graduate of the U.S. Military Academy, General Tuttle earned an M.B.A. from Harvard University.

Rayford B. Vaughn, Jr. is an associate professor of computer science at Mississippi State University, where he teaches graduate and undergraduate courses in software engineering and computer security. Previously, he was vice president, Military Integration Systems, for Electronic Data Systems (EDS) Corporation, McLean, Virginia, where he had full responsibility for all EDS contractual programs and support provided to the Defense Information Systems Agency and its customers.

Brian T. Wright is presently vice president of integrated architectures for Rockwell Collins. Prior to his present position, he was vice president of engineering for the Collins Avionics and Communications Division of Rockwell and vice president and director of engineering at ITT Aerospace and Communications Division. He has extensive experience in telecommunications, including circuit and message switching, secure voice processing, COMSEC equipment, and tactical combat net radios. Mr. Wright has a B.S. from Auburn University and an M.S. from the Naval Postgraduate School.

Appendix C

Meetings and Activities

FIRST MEETING

March 27–28, 2000
Holiday Inn Georgetown
Washington, D.C.

Embedded Software
Janos Sztipanovits
DARPA/ITO

Sponsor Perspective and Discussion
Donald Daniel
Deputy Assistant Secretary of the Air Force

A Vision for Weapon System Electronics Acquisition Through the Modular Open System Approach
Lt. Col. Glen Logan
Open Systems Joint Task Force (OSJTF)

Aging Avionics
Ellis Hitt
Battelle

Affordable Combat Avionics
C. Douglas Ebersole
Aeronautical Systems Center

Aging Avionics in Military Aircraft
Jim Johnson
Naval Air Systems Command

SECOND MEETING

May 2–3, 2000
Dayton Marriott Hotel
Dayton, Ohio

Open Systems Definitions
Butch Ardis
Aeronautical Systems Center

F-22 Avionics Program
Kenneth Fehr
Aeronautical Systems Center

F-15 Avionics Challenges
Lt. Col. Geoff Donatelli
Aeronautical Systems Center/FBA

Avionics Management Directorate Product Group Manager Perspective
Debby Walker
WR-ALC/LY

Affordable Combat Avionics
Lt. Gen. Robert Raggio
Aeronautical Systems Center/CC

Affordable Combat Avionics Initiative
Will Urschel
Aeronautical Systems Center/SMA

F-16 Combined Capability Improvement Program (CCIP)
Will Urschel
Aeronautical Systems Center/SMA

C-17 Affordable Avionics Road Map
Mark Wilson
Aeronautical Systems Center/YC

Weapon System Software Sustainment
Ajmel Dulai
Aeronautical Systems Center/EN

Open Avionics Systems Integration Study (OASIS)
Daniel Seal
Boeing Phantom Works

Systems, Technologies, Architectures and Acquisition Reform (STAAR) Report on Lockheed Martin's Proven Path Initiative
Tom Sarama
Lockheed Martin

Electronic Parts Obsolescence Initiative
Tony Bumbalough
Air Force Research Laboratory

Incremental Upgrade of Legacy Systems
David Corman
Boeing Phantom Works

Fiscal Policies/Funding Ground Rules
Donna Vogel
Aeronautical Systems Center/SMF

Affordable Avionics Solicitation Approach
C. Douglas Ebersole
Aeronautical Systems Center

Rapidly Transforming Electronics, a.k.a. Aging Avionics/Systems
Jon Ogg
Aeronautical Systems Center

THIRD MEETING

June 8–10, 2000
National Research Council
Washington, D.C.

Proven Path Program for Aging Avionics in Military Aircraft
Russell Frew
Lockheed Martin

Aging Avionics and the Army
Larry Johnston
Aviation Electronic Combat

Aging Avionics S&T Strategy and Program Status
Dale Johnson
Aviation & Missile Research, Development and Engineering Center

Component Obsolescence Management for Rotocraft Avionics Equipment
James Wasson
Smith Industries

Dynamically Adaptable Digital Architecture
Carl Abrahamson
TASC, Inc.

Bold Stroke
Daniel Seal
Boeing Phantom Works

JTRS
Roland Fritts
Raytheon

Synopsis of the DoD Evolutionary Acquisition Process
Noel Longuemare
Principal Deputy Under Secretary of Defense for Acquisition and Technology (retired)

Some Views on Aging Avionics, Modular Open Systems, and Related Issues
Noel Longuemare
Principal Deputy Under Secretary of Defense for Acquisition and Technology (retired)

FOURTH MEETING

August 17–18, 2000
National Research Council
Washington, D.C.

The Joint Strike Fighter
Maj. Gen. Michael Hough
JSF Program Office

Aerospace Industries
Brig. Gen. John Douglass
USAF (retired)

Avionics for Future Air Traffic Management
James Williams
Federal Aviation Administration

Rotocraft Center of Excellence and Rotocraft Industry Technology Association
Charles Morris
National Rotorcraft Technology Center

Color of Money
Blaise Durante
Assistant Secretary of the Air Force (Acquisition)

FIELD STUDIES

August 16, 2000
Andrews Air Force Base

Attendees: General Lyles, Robert L. Cattoi, William G.T. Tuttle, Jr., William C. Bowes, James E. Killian, and Noel Longuemare

August 4, 2000
Langley Air Force Base

Attendees: General Jumper, Robert L. Cattoi, William G.T. Tuttle, Jr., William C. Bowes, and James E. Killian

September 19, 2000
Pentagon

Attendees: Larry Delaney, Robert L. Cattoi, Noel Longuemare, James E. Killian, and Bruce Braun

November 30, 2000
Pentagon
Attendees: V.J. Garber, Robert L. Cattoi, William G.T. Tuttle, Jr., Noel Longuemare, Rocky J. Porzio, and James E. Killian